Grow Young
with
HGH

THE AMAZING MEDICALLY PROVEN PLAN TO

- Lose Fat, Build Muscle
- Reverse the Effects of Aging
- Strengthen the Immune System
- Improve Sexual Performance
- Lower Blood Pressure and Cholesterol

Dr. Ronald Klatz
WITH CAROL KAHN

HarperCollins*Publishers*

HarperCollins books may be purchased for educational, business, or sales promotional use. For information please write: Special Markets Department, HarperCollins Publishers, Inc., 10 East 53rd Street, New York, NY 10022.

FIRST EDITION

Designed by Alma Hochhauser Orenstein

Library of Congress Cataloging-in-Publication Data

Klatz, Ronald, 1955–
    Grow young with HGH : the amazing medically proven plan to
reverse aging / Ronald Klatz, with Carol Kahn. — 1st ed.
        p.   cm.
    Includes bibliographical references and index.
    ISBN 0-06-018682-8
    1. Somatotrophin—Physiological effect. 2. Longevity.   I. Kahn, Carol.
II. Title.
QP572.S6K53   1997
612.6'8—dc21   96-49609

97 98 99 00 01 ❖/RRD 10 9 8 7 6 5 4 3 2 1

GROW YOUNG WITH HGH

To the late DANIEL RUDMAN, M.D.
His vision and pioneering human research with growth hormone
for anti-aging marked the beginning of the end of aging,
and the birth of the "ageless society."

Contents

Acknowledgments

This book could never have been written without the help of the highly innovative growth hormone researchers, the doctors who pioneered in its clinical use, and the patients who shared their remarkable stories with us. There were far too many to name here and we do not wish to favor some over others, so we thank them all.

The authors gratefully acknowledge the enthusiasm and dedication of our agent, Jack Scovil, who was in on this project from the beginning, and our editor at HarperCollins, Vice President and Associate Publisher Gladys Justin Carr, for her insight and for not being deterred by controversy in our resolve to bring innovative, important, and lifesaving information to the public. We are also grateful to our co-editor, Cynthia Vail Barrett, for her astute editorial guidance. Special thanks to Dr. Steven Novil and Lisa Song for their invaluable research services and to Abigail Robin and Dee Pennisi for their unstinting help in preparing the manuscript. We also wish to thank Ali Orenstein and Debra Battjes for their work on the appendix. Carol is grateful to her husband, Ira, for his endless bounty of love and support. The authors also wish to acknowledge the generous help provided by the membership of the American Academy of Anti-Aging Medicine in uncovering numerous important research resources in this book. We especially wish to thank Robert Goldman, M.D., for the many hours that he volunteered in reviewing this text.

READER PLEASE NOTE:

Medicine is referred to as the "practice of medicine" because it requires constant reeducation and reevaluation to maintain proficiency and accuracy. The important information presented in this book includes research on digestion, nutrition, and anti-aging therapeutics written by the best and brightest minds and published by some of the most authoritative texts and journals in the world. But in less than five years from now we will know twice as much about anti-aging medicine and biomedical technology as we do today, and ten years from now we will have more than five times as much knowledge of this subject.

Because of the ever-expanding knowledge of medicine, the most any author can hope to do is wisely and prudently to put forth theory and practice as best as it is currently known. In preparation of this book, the authors reviewed thousands of published reports and hundreds of books, and interviewed many of the world's leading anti-aging researchers and scientists. This book is not intended to provide medical advice, nor is it to be used as a substitute for advice from your own physician.

If you wish to initiate any of the programs or therapies described in this book, *you must* consult a knowledgeable physician before doing so. Recruit him or her as your consultant in optimal health and longevity and, with your doctor's support, begin to use the ideas discussed in this book. By doing this, you will have taken the most prudent and powerful path to maximum life span and total well-being available via the new paradigm of anti-aging medicine. To find a physician who is a member of the American Academy of Anti-Aging Medicine (A⁴M) call (773) 528–4333. For updates and corrections regarding information in this book check our Website at http://www.worldhealth.net or call (312) 528–4333.

Introduction

In November 1993, in Cancun, Mexico, two events took place that had a profound effect on my life and formed the genesis of this book. One was the first scientific meeting of American Academy of Anti-Aging Medicine, which I had helped found and now serve as president.

The concept of the academy was a simple one but it represented nothing less than a seismic shift in medical practice. Throughout history, from the time of Hippocrates on, doctors have always drawn a clear distinction between disease and aging. Disease was a departure from health, a derailing from the normal track of life. Aging, on the other hand, *was* the natural course of events, as inevitable as the setting sun, as inescapable as gravity. Physicians attempted to treat the former and accept the latter. Now we had come together as a medical society to say that the process of aging was itself a disease. It was progressive and degenerative; it affected every cell, tissue, and organ in the body; and it was invariably fatal. The revolutionary principle on which our society was founded was that aging was amenable to treatment. There were interventions that actually worked. Some were already in clinical practice, others were being tested in patients, still others were being developed in the laboratory.

A great deal of excitement attended this first meeting as the men and women who had long cherished this dream with me talked about the possibility of putting the insights gleaned from basic and laboratory research into clinical practice. Anti-aging medicine—the slowing down, stopping, or even reversing the downward course of senescence—seemed to us an idea whose time had truly come. As we listened to scientific paper after paper detailing the progress being made in the new field of anti-aging medicine from antioxidant therapy to mind-body medicine to the discovery that turning off a single gene called "Age 1" in roundworms could increase their life span by 70 percent, we felt we were standing at the brink of a new era. Soon we would have an academy that would credential doctors who could practice the new medicine. We would set up a biomarkers program of aging that would establish for the first time standard measurements of how we aged. And through scientific meetings such as this one, we would present breakthrough research to audiences of thousands of physicians and scientists from across the planet—the leaders of the next generation of health and healing and anti-aging medicine.

Also present at the conference was another group of people, who were not part of the scientific crowd. They were a striking bunch, trim, muscular, sexy, energetic, upbeat, with the vibrant look of good health. Yet these were people in their fifties, sixties, even seventies. They talked about changes they had undergone—melting away of fat, increased muscle tone even in the absence of exercise, rapid healing from surgery, disappearance of wrinkles and cellulite, a vastly increased sense of well-being. They had come to Cancun not only to attend the conference but to receive a treatment that was not available at that time in the United States: growth hormone replacement. One woman scientist swore she could spot which men were on the drug by "a twinkle, no, rather a libidinous look in the eye."

Then a strange thing happened. The doctors and researchers who had assembled to form this professional organization devoted to aging were suddenly on the defensive. Could these claims be true? Had these people truly stepped back in time, or were they all suffering under a mass delusion? "My God," said one of the scientists, "we're all talking about the possibility of reversing the aging process and here these people are actually doing it."

Intrigued, I and my collaborator, science writer Carol Kahn, spent the next three years reviewing the literature on growth hormone, including more than 28,000 studies around the world; interviewing leading researchers in Sweden, Denmark, England, and the United States who have studied thousands of patients and conducted innumerable animal studies; and talking to physicians who have administered growth hormone and to many of their patients.

We have also researched the ways in which you can enhance your own stores of growth hormone through a program of diet, exercise, nutritional supplementation, common prescription drugs, and the latest method now undergoing clinical trials—secretagogues.

What we have learned is that basic and clinical research have consistently supported the idea that growth hormone replacement can reverse many of the aspects of aging. It now appears that the people we met in Cancun were neither self-deluded nor victims of hype. They were pioneers in the true sense of the word—opening the way for the rest of humanity, and in the process creating a new reality and new definitions of life, death, age, aging, mortality, and immortality.

PART 1

The Hormone
of Youth

No More Grumpy Old Men (or Women)

What if I told you could live to be age one hundred in as healthy and vigorous condition as you are now? Or if you are already suffering the ill effects of aging, you could turn back the clock twenty years and stay that way until the century mark? Would you be interested? And what if that extra thirty years added to the average life span allowed you to live long enough to take advantage of new scientific breakthroughs so that your healthy functional life span could be extended to 120 or 130 years and beyond? And what if the launching pad to that wonderful prospect involved taking a substance that caused your body to lose unwanted fat and build muscle, regrew your organs, improved your heart and lung activity, made studs out of men, increased sexual pleasure in women, and gave you the energy, sleep, and wonderful sense that the world is your oyster? Would you say, "What is this substance and how can I get some?"

There is such a substance. It is an Alice-in-Wonderland kind of hormone. Too little of it makes us dwarfs and too much of it turns us into giants. But the right amount of it at the right time promises to bring about the most fundamental revolution in society today— the beginning of the end of aging.

THE "FOUNTAIN OF YOUTH" HORMONE

On July 5, 1990, the prestigious *New England Journal of Medicine* published a clinical study on a drug that sent shock waves throughout the world. It was instantly hailed as a fountain of youth. In a scene that seemed like something out of the movie *Cocoon*, injections of synthetic human growth hormone—a substance naturally produced by the pituitary gland—had turned twelve men, ages sixty-one to eighty-one, with flabby, frail, fat-bulging bodies, into their sleeker, stronger, younger selves. In language rarely used in conservative medical journals, Daniel Rudman, M.D., and his colleagues at the Medical College of Wisconsin wrote: "The effects of six months of human growth hormone on lean body mass and adipose-tissue mass were equivalent in magnitude to the changes incurred during 10 to 20 years of aging."

In interviews with reporters, the men in the study and their wives reported other startling changes. The gray hair of a sixty-five-year-old man was turning black. The wife of another man had trouble keeping up with her newly energized husband even though she was fifteen years younger. A third man, who saw the wrinkles disappear on his face and hands, was now opening jars with ease, passing younger people on the street, and gardening for hours on end. Some of the users and their spouses made sly references to reinvigorated sex lives. What happened to the control group? In group 2, as they were called in the study, "there was no significant change in lean body mass, the mass of adipose tissue, skin thickness, or bone density during treatment." In other words, they continued aging on schedule.

DOUBLE-BLIND CLINICAL PROOF

Drugs and therapies that claim to reverse aging are nothing new. For decades youth seekers have trekked to the four corners of the world seeking youth and immortality. Recent examples include Ana Aslan's clinic in Rumania for injections of Gerovital H3, which is essentially the same Novocain your dentist uses. Konrad Adenauer, Gloria Swanson, Groucho Marx, and more than 50,000 lesser-known patients have had their buttocks shot with fresh fetal lamb cells at Clinique La Prairie in Switzerland. Others swear by ginseng or placental extract or nucleic acid therapy, which unleashed a craze for eating RNA-rich sardines. While some of these therapies might have genuine merit, none of them has passed the gold standard of drug testing: controlled, randomized, double-blind clinical study. Today there is immense interest in the hormone melatonin, released by the pineal gland in the brain. But melatonin has yet to be tested in double-blind trials in human beings. The only age-reversing drug that has passed placebo-controlled double-blind clinical trials with flying colors is human growth hormone. Not once, but many times over.

HGH TODAY

Following up his own work, Rudman found that HGH given to twenty-six elderly men regrew the livers, spleens, and muscles that had shrunken with age back to their youthful sizes. Improved muscle strength, he pointed out, could make the difference between someone's being on his feet or being confined to a wheelchair, between being spoon-fed or cooking a meal, between living independently or living in a nursing home. "The overall deterioration of the body that comes with growing old is not inevitable," he concluded. "We now realize that some aspects of it can be prevented or reversed."

Rudman's landmark discovery opened the floodgates to the point that there are now thousands of studies in the world medical literature documenting the benefits of growth hormone therapy. The National Institute on Aging has funded a multimillion-dollar effort in nine medical centers that will run for five years to test whether human growth hormone and other trophic factors—defined as substances that promote growth or maintenance of tissues—can reverse or retard the aging process. In 1992 medical researchers at Stanford University stated, "It is possible that physiologic growth hormone replacement might reverse or prevent some of the 'inevitable' sequelae of aging."

But many people are not waiting for the results of these studies. In this country and abroad, doctors at dozens of clinics are treating thousands of old and middle-aged people who wish to erase the effects of aging. Hundreds of case histories have now become available, some of which you will read about in the coming chapters.

As we go to press, *the FDA has just approved HGH for use in adults.* Until now the only indication had been for treatment of children who failed to grow due to lack of growth hormone. The adult indication is for somatotrophin (growth hormone) deficiency syndrome (SDS). Signs of SDS include decreased physical mobility, socialization and energy levels, along with a greater risk of cardiovascular disease and lower life expectancy.

Somatotrophin deficiency syndrome can have multiple causes, including pituitary disease, hypothalmic disease, surgery, radiation therapy, or injury. But by far the most common cause is aging, which, as this book makes clear, *is* a pituitary disease. The new FDA approval means that any doctor can now freely prescribe HGH to any patient with low levels of the hormone—that is, most people over the age of forty. According to Eli Lilly, the first drug company to receive approval for this indication, "The FDA cleared the new adult indication for Humatrope [Lilly's HGH product] after reviewing clinical data from multinational studies that were submitted to the federal agency by Lilly in August 1995." This included clinical trials which showed that "Humatrope [HGH] resulted in an increase in lean muscle mass, a decrease in body fat, an increase in exercise capacity and normalization of low HDL cholesterol levels" in adults with GH deficiency. Patients also reported "improvements in physical mobility and social isolation," on a standard quality-of-life questionnaire.

There is no doubt that other manufacturers of HGH (see list in Appendix) will apply for and receive permission to market the drug to adult patients. This competition will greatly reduce the price of the drug (which is already falling sharply), putting it within reach of most people who are concerned about their health and aging.

AGING IS NOT INEVITABLE!

Incredible though it may seem, the slow decline in appearance and function that we call aging may not be fixed in our genes or part of human destiny. In fact, as the elderly men in Rudman's experiment and the men and women we met at Cancun have shown, aging not only can be arrested but can actually be reversed. What we now call aging appears to be due in large part to the drastic decline of growth hormone in the body after adulthood. At age twenty-one, the normal level of circulating HGH is about 10 milligrams per deciliter of blood, while at age sixty-one it is 2 milligrams per deciliter—a decrease of 80 percent! It is growth hormone that grows the cells, bones, muscles, and organs of young children, and it is the falling levels of growth hormone after age thirty that slowly rob us of our youth.

By age seventy to eighty, according to Rudman, 38 percent of the population is as deficient in growth hormone as children who fail to grow normally because of their hormonal lack. But the prognosis for the failing aged person and the stunted child is equally good. As Rudman wrote in a 1991 article in *Hormone Research*, in terms of restoring hormonal levels, "the elderly hyposomatotropic [GH-deficient] subject is as responsive to human growth hormone as the GH-deficient child. Thus, the 'GH menopause' can be reversed by exogenous HGH."

Most of what we know about growth hormone in adulthood comes from research on people who are growth hormone–deficient because their pituitary glands have been damaged by disease or have been removed because of cancer. These people suffer a constellation of symptoms that look remarkably like aging. The results of growth hormone replacement in these patients, as shown in double-blind, placebo-controlled studies in many different countries around the world, have revealed that every feature associated with growth hormone deficiency can be reversed. In some cases, as you will see, the results were spectacular, with some patients going through changes almost as remarkable as those portrayed in the film *Awakenings*. But unlike those in the movie, the changes in these patients are permanent as long as they continue to take the hormone. As Bengt-Åke Bengtsson, a Swedish endocrinologist who is at the very center of this research, reported at an international workshop on adult HGH deficiency in Stockholm, "Growth hormone is considered by patients to be tremendously beneficial. Patients did not want to stop treatment. Moreover, nine out of ten commented on improved fatigue and physical performance and five out of ten also reported improved mood."

It is our contention that aging is, in large part, a pituitary deficiency disease (see Chapter 5). Growth hormone deficiency due to disease or removal of the gland is very rare, affecting about ten people per million annually. But insufficient growth hormone as a result of aging affects every single one of us.

The rejuvenating effects of growth hormone are global, acting on both mind and body, anatomy and physiology, form and function. In computer terms, it is like upgrading your microprocessor from a 286 to a Pentium chip. According to Allan Ahlschier, M.D., a family physician and radiologist, growth hormone replacement is the best anti-aging treatment money can buy.

HGH FOR MIND AND BODY

HGH is the ultimate anti-aging therapy. It affects almost every cell in the body, rejuvenating the skin and bones, regenerating the heart, liver, lungs, and kidneys, bringing organ and tissue function back to youthful levels. It is an anti-disease medicine that revitalizes the immune system, lowers the risk factors for heart attack and stroke, improves oxygen uptake in emphysema patients, and prevents osteoporosis. It is under investigation for a host of different diseases from osteoporosis to post-polio syndrome to AIDS. It is the most effective anti-obesity drug ever discovered, revving up the metabolism to youthful levels, resculpting the body by selectively reducing the fat in the waist, abdomen, hips, and thighs, and at the same time increasing muscle mass. It may be the most powerful aphrodisiac ever discovered, reviving flagging sexuality and potency in older men. It is cosmetic surgery in a bottle, smoothing out facial wrinkles; restoring the elasticity, thickness, and contours of youthful skin; reversing the loss of extracellular water that makes old people look like dried-up prunes. It has healing powers that close ulcerated wounds and regrow burned skin. It is the secret ingredient in the age-defying bodies of weight lifters and it enhances exercise performance, allowing you to do higher-intensity workouts of longer duration. It reverses the insomnia of later life, restoring the "slow wave" or deepest level of sleep. And it is a mood elevator, lifting the spirits along with the body, bringing back a zest for life that many people thought was lost forever. The latest research included in this book shows that it holds promise for the treatment of that most intractable and terrifying disease of aging—Alzheimer's.

The list of benefits seems to grow with each new study. They now include:

- 8.8 percent increase in muscle mass on average after six months, without exercise
- 14.4 percent loss of fat on average after six months, without dieting
- Higher energy level
- Enhanced sexual performance
- Regrowth of heart, liver, spleen, kidneys, and other organs that shrink with age
- Greater cardiac output
- Superior immune function
- Increased exercise performance
- Better kidney function
- Lowered blood pressure
- Improved cholesterol profile, with higher HDL and lower LDL
- Stronger bones
- Faster wound healing
- Younger, tighter, thicker skin
- Hair regrowth
- Wrinkle removal
- Elimination of cellulite
- Sharper vision
- Mood elevation
- Increased memory retention
- Improved sleep

HORMONE REPLACEMENT THERAPY

While growth hormone has an extraordinary range of age-reversing effects seen with no other substance, replacing hormones that decline with age is not new. The oldest continuing anti-aging experiment is hormone replacement therapy with estrogen or a combination of estrogen and progesterone in postmenopausal women. Premarin, a brand name of estrogen, is the single most prescribed drug in the United States. The benefits of female hormone replacement range from sharply decreased incidence of osteoporosis to vaginal lubrication to a 50 percent decrease in heart attack and strokes. A far smaller but growing number of men (and women also) are now taking testosterone to build muscles, increase strength, and recharge their sexual batteries. Two other hormones, DHEA (dehy-droepiandrosterone) and melatonin, which also decrease with age, have been shown to have broad anti-aging and anti-disease effects on the body. All these hormones should be part of an anti-aging program, as we discuss in Chapter 15. In fact, hormones play a controlling role in the "aging clock" as you will see in Chapter 5.

TWO BOOKS IN ONE

This book is to designed to be both an information guide to the age-reversing benefits of growth hormone and a practical, self-help workbook that will allow you to naturally enhance the levels of your own growth hormone. In the first section, we will tell you all about how growth hormone works in the body and why it is the master hormone of aging, and show how its replacement to the levels of young adulthood can actually restore your health, appearance, and function to what it was a decade or two earlier. You will meet some of the remarkable researchers who are carrying out clinical studies with growth hormone and the anti-aging physicians who use GH replacement in their practices, and hear the personal stories of people, including the doctors themselves, who have experienced firsthand its revitalizing effects. This section also includes the latest published and unpublished results from university-based scientists around the world on the effectiveness of growth hormone in treating a host of life-threatening ailments, such as AIDS, heart disease, stroke, osteoporosis, and emphysema. And you will learn why growth hormone may be the secret weapon in bolstering the immune system, maintaining brain function, and potentially extending the human life span.

In the second half of the book, we'll give you our Growth Hormone Enhancement Program. We'll show you how you can get all the benefits of growth hormone replacement on your own without spending up to $1,000 a month and taking twice-daily injections. This section will include everything you need to know about designing your personal age-reversal program. We'll give you a self-test so you can determine whether you need exogenous—or externally produced—growth hormone or can enhance your body's own stores. *Most important, we will provide a consumer's guide to all the nutrients and drugs that have been shown to release growth hormone in the body and tell you how to use safe, effective, natural substances to achieve your own rejuvenation.* We'll also show you how to maximize the effect with diet, exercise, and other hormone replacement therapy. And we'll advise on how to find a doctor to work with you, how to read your own lab reports like an endocrinologist, and how to avoid adverse side effects. We'll also bring you up to speed on a development that promises to change forever the way we age. These are the growth hormone secretagogues that stimulate the pituitary to release its own stores of natural growth hormone. Clinical studies now being conducted by major pharmaceutical houses on secretagogues indicate that one pill a day may be as effective as—or even more effective than—growth hormone injections in bringing back the surging growth hormone levels associated with peak bodily functioning.

EVERYONE CAN BENEFIT

Growth hormone therapy is not just for those who are grossly deficient in the hormone. As Dr. Rudman and his colleagues pointed out in their paper, "in middle and late adulthood *all people* [emphasis added] experience a series of progressive alterations in body compositions," including a loss of lean body mass, and increase in fat tissues, and atrophy of skeletal muscle, liver, kidney, spleen, skin, and bone. "These structural changes have been considered unavoidable results of aging," they wrote. "It has recently been proposed, however, that reduced availability of growth hormone in late adulthood may contribute to such changes." In other words, everything we associate with aging—from middle-aged spread to shrunken, bent-over, frail, doddering senescence—may be due wholly, or in part, to the decline of growth hormone.

There is no reason to wait until you are a senior citizen to enjoy the benefits of HGH replacement. In fact there is every reason to believe it is better to start at a much younger age when the levels of your own hormone have already begun to decline. Dr. Eve Van Cauter, a human growth hormone researcher at the University of Chicago Medical Center, says, "All of these ideas about treating people with growth hormone have been directed toward people 65 and older. If you look at the data, people have so-called 'elderly' levels by age 40. Perhaps we should be giving human growth hormone replacement therapy earlier rather than attempting to treat tissues that have seen little or no growth hormone for decades."

If you are like one-third of the older population and are severely deficient in growth hormone, then HGH therapy with injections of the hormone is for you. We will tell you how to determine if you are in that group, how to obtain a supply of the hormone, and how to locate doctors skilled in its use. But if, like two-thirds of the over–forty-five population, you are still making small, if declining, amounts of growth hormone, *you can achieve the same effect at a fraction of the cost using prescription drugs or nutrients from the health food store to stimulate greater release of your endogenous—internally produced—hormone.* And you can do it safely, effectively, and economically.

SHOULD YOU TAKE HGH?

After the enormous positive publicity after the Rudman study, articles began appearing warning about the effects of human growth hormone. Much of the concern centered on the abuse of HGH by athletes to enhance their performance. Massive doses of growth hormone taken by athletes bent on attaining superhuman size and strength led to acromegaly—the overgrowth of many bodily parts. Acromegaly is a dangerous disease with increased incidence of heart attack and death. Other side effects reported in the Rudman study, such as water retention and carpal tunnel syndrome, were also worrisome, but these effects were also due to large doses, often four times that which was needed to obtain the desired benefits. A recent three-year study of growth hormone treatment in Denmark using lower doses found no side effects. And in a 1995 review paper on the use of human growth hormone in hormone-deficient adults, Drs. Rosén, Johannsson, Johannsson, and Bengtsson of the University Hospital of Göteborg, Sweden, concluded: "There is no evidence suggesting that this [growth hormone] replacement therapy causes any unfavorable long term side effects." More recently, Dr. George Merriam of the University of Washington in Seattle, who is conducting one of the National Institute on Aging studies on growth hormone–releasing hormone, or GHRH, was quoted in the *New York Times*, July 18, 1995, as saying that the preliminary findings of his team includes "a complete absence of side effects."

THE EXPANDING HUMAN LIFE SPAN

There are other more subtle reasons that some writers in popular magazines have criticized growth hormone. It is always scary venturing into uncharted territory. Many people believe that aging is "natural" and preventing it goes against "nature" or a divine plan. But the natural history of human beings is one of increased survival and continued conquest over disease.

The life expectancy of human beings has tripled since Roman times, and since the turn of the century alone, we have added another twenty-five years to the average life span. By the end of the nineteenth century, the advent of vaccines made the first inroads against infectious disease and the "silver bullet," salvarsan, became the first weapon specifically targeted against a disease—syphilis. With advances in antibiotics and surgical techniques, the death rate from battlefield wounds during the Vietnam war was reduced by almost two-thirds from that of World War I. Now we are beginning to conquer the devastating modern-day killers of heart attack, cancer, and stroke. These are the so-called diseases of aging, because you have to live long enough for them to make their appearance.

Silently, with almost no recognition by the mass media, a life extension revolution has been taking place over the last few decades. Catching up, the *New York Times* trumpeted in a page one story on February 27, 1996, "New Era of Robust Elderly Belies the Fears of Scientists." According to the article, the National Long Term Care Surveys, which follows people age sixty-five and over enrolled in Medicare, had found that since the survey began in 1982, the percentage of people requiring care had fallen at a rate of 1 to 2 percent a year! The main reason was simply that seniors were less sick. For instance, in the seven-year period from 1982 to 1989 alone, there was a dramatic decline in the number of people over sixty-five with high blood pressure, arthritis, and emphysema. The survey estimated that compared with the disability rates of 1992, there were 121,000 fewer disabled people in 1995. With less disease, the death rates from killing disease plummeted, with mortality from strokes in the period from 1960 to 1990 falling 62 percent, while that of heart disease dropped 46.9 percent.

People over eighty-five now form the fastest growing segment of the population. Even reaching one hundred is no longer considered remarkable, and 121 years, as recently celebrated by Madame Jeanne Calment in France on February 21, 1996, seems well within reach of many of us alive today.

My own prediction is that this long-lived trend will not only continue but accelerate. If our present scientific knowledge base continues to double every 3.5 years, we will have a sixteen-fold increase in basic knowledge by the year 2010. These medical and technological advances will allow us to do things that are almost beyond imagination today. Until recently, life spans of one hundred years seemed like science fiction. But in the next thirty years life spans of 120 to 130 years will not be uncommon, with these ultra-long-lived people in a state of sound mental and physical function. In the final chapter, I will discuss the predictions of American Academy of Anti-Aging Medicine of the advances we can expect to see in the next five, ten, thirty, and fifty-five years that will lead to a life span so far beyond that of today, it might be considered "human immortality."

LETTING NATURE TAKE ITS COURSE

In the movie *Grumpy Old Men*, Jack Lemmon and Walter Matthau portray two bickering sourpusses, beset by aches and pains, winding down with age. These men are on the opposite end from the youthful, athletic, even dashing seniors shown in TV commercials aimed at the Social Security crowd. Yet they are far closer to many of the doddering elderly people we see on the street, in our families, and in nursing homes.

What happened to the Lemmon and Matthau characters is what happened to the control group in Rudman's study. It is the destiny of us all if we let nature take its course. Dietary modifications and exercise can delay or prevent some of the decrements of aging, but actual age-reversal requires putting back what was lost over the years. If we are to live long and prosper, we need to avail ourselves of substances that will make our later years as vital and meaningful as our earlier ones.

IT'S YOUR CHOICE

Grumpy Old Men points the way to the past, not to the future that is already here. The path it shows is a path that we no longer need travel. If you are in your thirties, taking natural substances that stimulate growth hormone release can keep your own bodily supply from decreasing. If you are in your forties or fifties, or older, our program of growth hormone enhancement can recharge your dwindling supply. If you have ceased to manufacture the hormone altogether, growth hormone replacement can give you back what nature has taken away. It can literally give you a new lease on life, granting you tenancy in a world you thought was lost to you forever.

By replenishing your supply of growth hormone, you can recover your vigor, health, looks, and sexuality. For the first time in human history, we can intervene in the aging process, restore many aspects of youth, resist disease, substantially improve the quality of life, perhaps even extend the life span itself. The "Fountain of Youth" lies within the cells of each of us. All you need to do is release it.

The Story of Growth Hormone

Americans like things that come in large sizes—cars, houses, laundry soap, men. Our idols are oversized athletes, hulking football players, muscular basketball giants. In the personal ads, "tall" is the single most desirable physical attribute for men. Stature is equated with leadership, whether it is the head of a corporation or the President of the United States. A song that tweaked our hidden prejudices about height went straight to the top of the charts: "Short people got no reason to live."

So at first it seemed as though growth hormone was the miracle answer to the problem of children whose bodies made too little of the precious hormone. Left untreated, they were doomed to being abnormally short or even dwarfed. But the story of growth hormone's development as a drug was far from smooth. It encompassed both triumph and tragedy, the greatest pitfalls and the most far-reaching promise.

THE CADAVER CATASTROPHE

Although the hormone was discovered in the 1920s, it was not until 1958 that a pioneering endocrinologist at the New England Medical Center in Boston, Maurice Raben, first injected it into a growth-stunted child whose body produced none of the hormone. It worked. The child began to grow. Other doctors soon followed suit, and the treatment of growth hormone–deficient youngsters became a reality.

The source of the human growth hormone was the only one available—the brains of cadavers. It took thousands of dead brains to obtain the few tiny drops of the hormone that could be injected into the children's tissue. Most of the cadaver brains came from Africa and were shipped to the factories of commercial drug manufacturers, who would extract the hormone from the pituitary gland. Since heating the hormone would destroy it, the manufacturers sterilized the extract through a kind of pasteurization. But by the 1980s, when thousands of children were receiving the treatment during their growing years, it became evident that the hormone which promised them a normal life would end up robbing some of them of a chance to live it.

Creutzfeldt-Jakob disease (CJD) is a horrendous disease, characterized by a progressive dementia and loss of muscle control, usually killing its victim within five years after the onset of symptoms. (It is also called mad cow disease in England, where it has been linked to eating beef from infected cattle.) When three children who were taking growth hormone extracts developed this same rare disease, which generally affects about one person in a million, the conclusion was unmistakable. The deadly virus had hitchhiked onto some of the hormone. The FDA ordered the distribution of the drug stopped.

By 1991 seven children in a group of 5,000 hormone recipients that had been studied had developed CJD in the United States, *with fifty cases worldwide linked to injections of pituitary-growth hormone.* But researchers warn that since the disease, like AIDS, is caused by an infectious agent that can take up to fifteen years to cause symptoms, it is too soon to know how many children will fall victim. Only one-tenth of the group has been followed that long.

RECOMBINANT GH IS BORN

With the cadaver source lost, the drug had to be synthesized from scratch. This was a monumentally difficult task.

Hormones are proteins and proteins are made of building blocks known as amino acids. Growth hormone is the largest protein produced by the anterior pituitary gland body with 191 amino acids. Fashioning a molecule that size was as delicate and difficult as putting together a chandelier with 191 chains of lead crystal, only in this case you had to identify what each chain was and where it hung.

But as luck would have it, the 1980s were a particularly propitious time to develop a growth hormone drug. First, in 1985, Congress passed the Orphan Drug Act as a financial incentive to develop drugs for rare conditions that affected fewer than 200,000 patients. In return for the cost of developing these orphan drugs, drug companies would be given seven years' exclusivity; no one else could enter the market. Second, the new technology of genetic engineering had just come on board. Gene splicing allowed scientists to clone the proteins that make up the human body. This meant identifying the exact sequence of the DNA for that particular protein. Once the protein was cloned, the reproductive machinery of a common bacterium found in the human gut, *E. coli*, could be harnessed to turn out an endless supply of the drug.

Human growth hormone was just what Herbert Boyer, the Nobel prize–winning scientist who helped invent genetic engineering, had in mind when he co-founded Genentech in 1976. He put his best man on the job, David Goeddel, a shy, soft-spoken scientist who wore his creed on his T-shirt: "Clone or Die." Boyer knew he was in a race with the drug manufacturing giants for the winner-take-all market. A year later Goeddel and his team were racing to the courtyard of Genentech, where they cupped their hands together like a megaphone and howled the magic letters, DNA. They were signaling their fellow employees in the peculiar ritual of their company that they had successfully completed a cloning. In 1985 Genentech, the same company that earlier had successfully cloned the human gene for insulin, brought out the second recombinant DNA drug ever developed.

But their triumph was short-lived. The biosynthetic hormone, which they named Protropin, differed from its human counterpart by one amino acid. Although this in no way affected its performance in the body, this tiny variation opened the door to competition. The following year, the Indianapolis-based drug company Eli Lilly succeeded in making a 191-amino acid growth hormone that was 100 percent identical, physically, chemically, and biologically, to the one made by the human pituitary. They too applied for the protection of the new drug, which they called Humatrope, under the Orphan Drug Act. A court battle ensued in which Genentech lawyers argued that since their company was the first to come up with a synthetic growth hormone, it was clearly the victor under the Orphan Drug Act. Genentech may have been the first in time, countered Eli Lilly's lawyers, but it wasn't the first in quality. That distinction belonged to Lilly. In the end, both companies were allowed to manufacture and distribute the drug, with all other companies shut out of the U.S. market. With a year's supply of the drug costing a patient about $14,000 to $30,000, the nation's growth hormone–deficient children still brought the two companies more than $175 million. Not bad for an orphan. But it soon became clear that the real potential of the drug lay in what it could do at the other end of the life cycle.

DR. RUDMAN'S TRIUMPH

In Madison, Wisconsin, an endocrinologist by the name of Daniel Rudman finally got the break he was waiting for. Rudman had long been intrigued by the idea that the decline in hormonal activity may dictate the slow deterioration of the body starting around age thirty-five. The only way to test this was to replace the missing hormones in elderly individuals and see if he could reverse some of the changes associated with aging. He decided to start with growth hormone.

He based his choice on two reasons. First, he knew that the decline of GH after age thirty-five was often accompanied by changes in the body composition. After age thirty-five, the average weight of a healthy man changes very little, but the body undergoes a shift comparable to a firm young McIntosh apple turning into a soft, mushy one. The amount of body fat expands by 50 percent, while the lean body mass (LBM) that forms the muscles, bones, and all the vital organs shrinks by 30 percent. This means that the structure of the body is crumbling at the same the functional capacity is declining. Like an old car, our bodies are rusting out and running down.

Second, researchers in Denmark and Sweden had found that the recombinant growth hormone restored a lean body contour in both children and adults with growth hormone deficiency due to pituitary problems. It also seemed incredibly safe, since more than thirty years' experience in children had failed to turn up any significant side effects associated with its use. But no one had yet shown whether the hormone might have a similar benefit in healthy older people.

In a remarkably prescient paper in the *Journal of the American Geriatrics Society* in 1985, "Growth Hormone, Body Composition, and Aging," Rudman advanced the idea for the first time that growth hormone could reverse aging. First he summarized the mechanisms for the loss of structure and function in old age: the "irreversible effect" of the aging process itself caused by the accumulation of unrepaired damage to the DNA; loss of cell division; errors in protein synthesis; the cumulative effects of diseases like atherosclerosis, hypertension, infections, and autoimmune disease; the effects of physical inactivity or disuse; and, in women, the menopause. Then he wrote, "We now propose an additional novel mechanism for geriatric regressions operative in about half of the elderly population: the cessation of endogenous GH secretion."

He decided to test his hypothesis by looking at the changes in body composition with age. If growth hormone replacement could reverse these changes in body composition, Rudman reasoned, it might affect the loss of structure and function that accompanies aging. The study group consisted of twenty-six men between the ages of sixty-one and eighty who had been recruited through newspaper advertisements. They all showed the characteristic pear shape of aging, having lost the firm, muscular definition of youth. They were all severely deficient in growth hormone as shown by testing but were apparently otherwise healthy. In the end, the study followed twenty-one men, four having dropped out for personal reasons and one having been found to have prostate cancer. Twelve men received injections of HGH three times a week, while the nine men in the control group received no treatment.

At the end of six months, the difference between the two groups was startling. The men had been told at the start of the study not to alter their lifestyle, even if it included smoking, drinking, and lack of exercise. Yet those who had been on the drug gained an average of 8.8 percent in lean body mass and lost 14 percent in fat, without diet or exercise! Their skin had become thicker and firmer, and the lumbar bones of the spine had increased in density. As noted earlier, in the report in the *New England Journal of Medicine,* Rudman and his colleagues called the changes in body composition "equivalent in magnitude to the changes incurred during 10 to 20 years of aging."

Rudman's experiment was immediately hailed as the breakthrough it was. The invincibility of the aging process had been shattered once and for all. Like the first people to have received cow pox vaccine, or the first heart transplant recipients, these twelve men had changed the course of history forever.

While the benefits were clearly striking, there were some uncomfortable side effects, including the painful wrist condition known as carpal tunnel syndrome, and gynecomastia, or enlarged breasts. The problem, as Rudman himself was now aware, was that the dosages he used were too large. "The results of our clinical trials with HGH have revealed the optimum hormone dose is only one quarter to one half as great as was previously believed," he wrote. At the lower doses, he believed, all the beneficial effects should be retained without any adverse side effects. Far from being discouraged, Rudman was looking forward to new experiments to see if GH could significantly improve the quality of life in the frail elderly. Tragically he was not to have that opportunity. Stricken with a pulmonary embolism, he died in 1994 at age sixty-seven. But his legacy is an everlasting one. He inspired physicians and scientists all over the world to look at the possibility of growth hormone replacement in the treatment of aging and age-associated diseases. And he proved conclusively that what we call aging is neither inevitable nor irreversible.

At the same time that Dr. Rudman was beginning his research on growth hormone replacement in Wisconsin, across the Atlantic researchers in England, Denmark, and Sweden were starting theirs. Unlike their American counterpart, the European researchers were looking at people whose lack of growth hormone was due to disease. The groups were headed by Professor Peter Sönksen at St. Thomas' Hospital Medical School in London, Dr. Jens Sandahl Christiansen at the University of Aarhus in Denmark, and Dr. Bengt-Åke Bengtsson at the University of Göteborg in Sweden. They all found what Sönksen called "remarkably consistent results."

THE PATIENTS OF SAHLGRENSKA

Dr. Bengtsson, an exuberant endocrinologist who always appears to be bouncing from one section to the next of his large research and clinical setting at the University of Göteborg's Sahlgrenska Hospital in Sweden, is a man who clearly relishes a challenge. When he first began his work in the mid–1980s, the idea that growth hormone played a role in anything other than the skeletal growth of children was not a popular one. "If you look at the textbooks of endocrinology you can read that GH has no effect in the adults," he says. "So the idea to treat adults was a little crazy."

But there were a few disturbing hints in the scientific literature that this might not be the case. In the 1960s hypophysectomy, or the removal of the pituitary gland, was used as a treatment for breast cancer, diabetes, and acromegaly. The patients were given routine replacement of corticosteroids, thyroxin, and sex steroids, but not growth hormone, which was in short supply, and which was deemed unnecessary anyway. But in 1962 Raben, who had already won acclaim for increasing growth in a pituitary dwarf with human growth hormone, decided to try HGH in a thirty-five-year-old woman who had been on conventional replacement therapy for eight years. After two months, he noted that she had increased vigor, ambition, and sense of well-being.

A year later Swedish scientist Thomas Falkheden described in his doctoral thesis the outcomes of patients who had their pituitary glands removed. Within a month after their operation, they had lower basal metabolism, reduced blood volume, and impaired heart and kidney function despite hormone replacement for the adrenal, thyroid, and gonadal hormones. He hypothesized that the reason for the detrimental changes in these patients was the missing growth hormone. But because growth hormone was in such short supply, nothing was done to follow up this work until the recombinant growth hormone became available.

Bengtsson began to wonder what happened to those patients whose growth hormone was not replaced. He asked some of his senior friends on the hospital staff. "They are doing perfectly well," they told him. "No problem whatsoever." But one physician said, "They never come back." The few that did show up in Bengtsson's outpatient clinic were not encouraging. "They looked like zombies," he said. "They moved in slow motion."

One patient, in particular, who concerned him was a man who had had a good career in the computer business until he developed a pituitary tumor. After the removal of his gland and even with replacement hormone therapy, his career went downhill. Finally he was no longer able to keep his job and was placed on disability.

Bengtsson decided to add growth hormone to the man's hormonal replacement. Fortunately a Stockholm representative of the American drug company Eli Lilly, one of the two manufacturers of human growth hormone, supported the idea. Bengtsson flew to Indianapolis to meet with the executives of the drug company and came away with a commitment of $150,000 to get his research under way.

In the end Bengtsson carried out two lines of research. One was a follow-up of all those patients who had been treated with pituitary gland removal and the other was the replacement of growth hormone in a number of people who were growth hormone–deficient because of pituitary problems. The two studies fit together like hand in glove and, along with the findings of his colleagues in England and Denmark, turned the knowledge about growth hormone in adult life on its ear. These revelations are still continuing, as we will see in the coming chapters.

First was the problem of what happened to the patients whose pituitary glands had been removed. Why were many of them lost to follow-up in the land of national health care? There were 333 patients who had been diagnosed with pituitary insufficiency in a thirty-year period between 1956 and 1987. All these patients had been treated with pituitary hormone replacement, including cortisone, thyroid hormones, and sex hormones. The one hormone that was not replaced was growth hormone. The answer appeared to be that they had died prematurely as a result of their growth hormone deficiency, at twice the expected rate—107 deaths compared with 57 in the overall population matched for age and sex.

The major reason for the dramatic rise in mortality was an almost twofold increase in cardiovascular disease, with sixty deaths versus thirty-one in the general population. "This surprised us all," says Bengtsson. "It was the first study ever to show the long term prognosis of pituitary deficiency with conventional replacement therapy. And it indicated that growth hormone deficiency was associated with atherosclerosis."

He and his student Thord Rosén then went on to find that these patients had a number of other problems. These included increased fat mass, with most of the fat concentrated in the gut; decreased muscle mass; reduced bone mineral content; twice the fracture rate of the general population; problems in thinking and remembering; and, most striking of all, a poor quality of life. They complained of fatigue, low vitality, low self-esteem, social isolation, and an impoverished sex life, and they were far more apt to be on disability pension than their fellow workers.

"When you look at quality of life, you must realize that most physicians are poor in detecting problems with it," says Bengtsson. "We ask specific questions related to things like blood pressure or the heart, but we don't ask, 'how do you feel?'" But the scores these patients were getting on the Nottingham Health Profile, a self-assessment questionnaire on the quality of life, indicated that many of them were struggling with problems of low self-esteem, anxiety, and depression. The psychological profile gave a striking vindication to the "zombie effect" that Bengtsson had observed in his clinic.

Bengtsson and his colleagues at Sahlgrenska began putting a group of these patients with pituitary deficiency on growth hormone. They didn't have to wait long for results. "We called it the Lazarus effect because it was like that," he says, snapping his fingers. "We woke them up. Some, not all. With some patients it was like giving them a kick in the back. Their lives changed within a few weeks." In the end, the treatment benefited everyone who participated in the program. According to Dr. Lena Wiren, a psychologist who evaluated the patients, "Nobody wants to stop treatment. Sometimes it is not even the patient who notices the difference it is making. Rather it is their wives or children or friends at work."

In some cases the changes amounted to a personality transplant. She recalls one patient who slept sixteen to eighteen hours a day. When he was awake, his friends wished he would go back to sleep since he was angry all the time. He spent many of his waking hours phoning his friends to apologize for having been so angry the day before. Today, three years later, this patient has had a complete turnaround. Instead of sleeping sixteen hours, he works that amount of time in his own restaurant, which has become one of the most famous in Göteborg. His happiness is so great that recently he invited Dr. Bengtsson and his staff to his restaurant to celebrate his amazing recovery.

Another case concerned a lonely, fat young man of twenty. Lacking a job, he spent all his spare time in bed reading children's comic books. He came to Sahlgrenska Hospital hoping to be certified for a disability pension. Instead the staff talked him into treatment with growth hormone. He now has a girlfriend and a job, and he spent the last summer traveling through Europe by train. But his greatest pleasure occurred when he went to buy liquor in a store in Sweden. When he presented his ID, the salespeople balked. "No," they said, "this photo isn't you." The young man showed Dr. Wiren the card and she immediately understood. The picture on the ID was of a sullen, fat, baby-faced young man. Now he was slim and the look on his face was one of happiness. "They're right," she said. "It isn't you."

ANSWERS TO AGING—JULY 1, 1998

The story of growth hormone is still being written. As we mentioned in Chapter 1, Rudman's groundbreaking study inspired the National Institute on Aging in 1992 to conduct nine clinical trials on so-called trophic factors that will take five years to complete. Six of these focus specifically on growth hormone or growth hormone–releasing hormone. Marc Blackman, M.D., of the Johns Hopkins University Bayview Medical Center is looking at GH and sex steroids in restoring a number of aspects that decline with age (see below for details); Dr. Andrew Hoffman of Palo Alto Veterans Administration Medical Center and Stanford University Medical Center is testing the metabolic effects of GH and a related hormone, insulin-like growth factor-1 (IGF–1) in healthy older women and also studying whether these substances might enhance the effects of an exercise program; Robert Schwartz, M.D., of the University of Washington Harborview Medical Center is evaluating the effects in older women of growth hormone–releasing hormone when used alone or in combination with endurance or strength training; David MacLean, M.D., of Rhode Island Hospital is testing whether GH and exercise have a synergistic action in improving muscle performance in older men and women; and Stephen Welle, Ph.D., at the University of Rochester School of Medicine and Dentistry is examining the mechanism of muscle atrophy, using strength training, growth hormone injections, or both on muscle atrophy. Two other studies, by Joyce Tenover, M.D., Ph.D., at Emory University School of Medicine, and by Peter J. Snyder at the University of Pennsylvania School of Medicine, are seeing whether testosterone replacement in older men improves age-related declines in muscle mass, strength, bone density, and cognitive function. Yet another study, by Michael Kleerekoper, M.D., is looking at the effectiveness of hormone replacement therapy in women in preventing bone loss in the lower leg that is usually involved in hip fracture.

These are unprecedented, potentially historic, studies because they are not looking at the effects of aging in fruit flies, worms, fish, mice, and rats, but are examining actual interventions in elderly men and women. And unlike studies at the other National Institutes of Health, which focus on specific various diseases, these will zero in on the effects of the aging process itself. As NIA director Richard J. Hodes writes, "The National Institute on Aging's mandate is to conduct research with the aim of improving the quality of life and maintaining the independence and vitality of people well into their later years." The results of these studies should go a long way to fulfilling that mandate.

One of the most ambitious studies is being carried out by Dr. Marc Blackman, chief of endocrinology and metabolism at Johns Hopkins Bayview Medical Center. He is looking at four groups of healthy men and women over age sixty-five—the oldest person enrolled so far is eighty-three—who will receive growth hormone with or without testosterone (in men), GH with or without estrogen and progesterone (in women), or control groups that receive placebo treatment (both men and women.) The treated groups will have hormone levels raised to that of a thirty- to forty-year-old person.

The scope of what Dr. Blackman and his associates are looking at is dazzling. While most clinical research is content to look at changes in one or two parameters, such as a rise or fall in cholesterol levels, "we are doing a monumental series of measurements that is, I believe, more than anyone, anywhere, has done," says Dr. Blackman. At the beginning and end of the treatment, the patient stays in the hospital for three days and goes through an "exhaustive battery of physical, psychological, functional, biochemical, radiologic, and molecular tests designed to highlight areas like body composition, specifically decreases in the amount and strength of muscle, the amount and strength of bone, and increase in body fat especially abdominal fat, changes in metabolism such as abnormal sugar metabolism with an tendency to diabetes and abnormalities in cholesterol metabolism, heart function, lung function, kidney function, immune function, skin, behavior and mood, and even that nebulous, but all important, quality-of-life issue."

All the studies are double-blind so that nobody, not even the investigators, knows who is getting what. "Obviously, everybody wants to know, the subjects want to know, the media want to know, the investigators want to know, but we can't tell them. We won't know until midnight, July 1, 1998, when the studies are slated to be completed and the codes telling us which are the treatment groups and which are the controls are decoded."

But Blackman, who calls himself an eternal optimist, expects that the studies will bear out Rudman's findings of age-reversal among elderly people given growth hormone replacement. "They are likely not only to confirm, but to go far beyond what Rudman did," he says. The work of Bengtsson in Sweden, Sönksen in England, and Christiansen in Denmark on adults with GH deficiency as a result of pituitary disease "is a model for what we are doing in aging," says Blackman. "Their work clearly shows that growth hormone deficiency has all these adverse consequences in the systems we're looking at. We are making the hypothesis that to a lesser degree, the age-related declines do the same thing as the deficiencies due to disease and that careful supplementation in an effective and safe way will improve the situation in regard to aging."

With the NIA studies and the voluminous research on aging and growth hormone and hormonal replacement, says Blackman, "over the next few years, we are going to get a lot of information and over the next five or ten years there will be an explosion and beyond that, you can't even foresee," he says. "I really believe there will be an impact not just in understanding but an impact that will translate into improved well-being of older people. If the aggregate work that is being done all over the world is successful, it will be very satisfying to prolong function and to retard muscular, skeletal frailty and the loss of physical and psychological independence."

WAIT OR ACT?

Should you wait until all the results of these studies are in before deciding to do growth hormone replacement therapy? Or should you act now to obtain these benefits for yourself? According to Dr. Stanley Slater, deputy associate director of the NIA for geriatrics, after the results are decoded in 1998, "results will be analyzed over the following year. By the mid to end of 1999, we should have some publications." These will be followed up, no doubt, by more studies designed to answer new questions that are raised by these studies. It took forty years before the medical establishment gave its nod for routine replacement of estrogen and progesterone in post-menopausal women, and it might take another forty years before it gives the nod to growth hormone replacement.

We believe that the consequences of *not acting* are far worse than the consequences of acting. The aging of the body is not a benign process. It takes its toll in accumulated loss of physical and mental function, increasing vulnerability to sickness, and devastating blows to our self-image as attractive, vital, sexual beings. And it happens every day in every cell in our bodies.

This book is designed to give you all the facts you need to make up your mind about something that literally has life-or-death consequences. Dr. Thierry Hertoghe, a physician specializing in hormone replacement therapy in Brussels, points out that the 333 growth hormone–deficient adults in Bengtsson's study had double the risk of dying prematurely. "But an adult of fifty typically has the same production of growth hormone as a young adult who is growth hormone deficient," says Hertoghe, "so he has twice the chance of dying from heart disease as a person who is treated with growth hormone. I find it sad that people are aging. They complain, they have a lesser quality of life, they die prematurely. All these people could benefit from hormonal treatment. When you see all the good effects, you say, 'Why isn't everybody in treatment?' This is the medicine of the future."

What is true of the patients of Sahlgrenska Hospital is true of all of us. Before growth hormone therapy, these GH-deficient people, some of whom were in their twenties and thirties and whose average age was mid–forties, provided an almost perfect picture of the frail elderly. They had the body contours of Rudman's old men, failing hearts, fragile bones, low energy levels, and poor outlook on life, and were apt to be semi or fully retired. After treatment, every single one of these traits had been reversed. The same dramatic, even Lazarus-like effect has occurred in people who have undergone growth hormone replacement in this country. But in this case it is not people with pituitary diseases who are being treated, it is doctors, lawyers, business executives, screen stars, anyone with the money and the determination not to grow old.

In the next chapter we will look at the first clinical results of more than 800 of these men and women who have been treated with growth hormone replacement for more than six months. They are among the thousands of other older people who are not waiting for the medical establishment to come on board. They feel that they literally do not have the time. And as we will show in the second half of this book, with the Growth Hormone Enhancement Program, you may be able to secure the same extraordinary benefits at a fraction of the cost.

GH Therapy in 202 Aging Adults

BACKGROUND

The largest clinical study on growth hormone replacement began with the troubles of one man. In 1992, at the age of forty-two, Dr. Edmund Chein, a rehabilitation specialist by training, felt as though he was rapidly going downhill. He was gaining weight and already had a huge pot belly, his cardiovascular risk factors like elevated cholesterol and triglycerides were in the danger zone, and worst of all he had started to have chest pain. His cardiologist confirmed his worst fears, telling him that his coronary artery was blocked and that any day he might have a heart attack. He immediately suggested that Chein go on cholesterol-lowering medication.

Chein decided to try a far less conventional approach. As a physical medicine and rehabilitation medicine specialist, he had used hormone replacement therapy in people whose endocrine glands had been damaged as a result of injury. He had watched as people given testosterone and human growth hormone regained function without adverse side effects. He had also read Dr. Rudman's study in 1990 with tremendous interest and stored it in the back of his head.

"Other things like melatonin and DHEA had been shown to slow aging. But nothing on the planet had been shown to reverse aging until Rudman showed that it could. I decided to try growth hormone replacement on myself. I wanted to replace all my hormones to that of a 20-year-old and see if I could reverse my biological age. My theory was if I could truly revert to the point that I was at age 20, then I shouldn't have any more symptoms since I didn't have any when I was that age. My cholesterol should come down, my triglycerides should come down, my chest pains should go away."

He went to a fellow physician who found that his levels of growth hormone, testosterone, thyroid hormone, and DHEA were all low. Chein replaced all four hormones to the point of a twenty-year-old and within six months, Chein had his answer. "My chest pains stopped, my cholesterol and triglycerides become normal, and I started losing my big belly." Chein's doctor was so impressed that he decided to put himself on hormone replacement therapy. Soon he was noticing the same "miraculous changes" in himself, according to Chein.

When the El Dorado clinic opened in Cancun, Mexico, Chein wondered if he could open a clinic in the United States. El Dorado had been set up in Mexico to avoid the FDA's ban on the use of human growth hormone on adult patients that was in effect before August 1996. Chein, who is also a lawyer, believed that its use in the United States was perfectly legal. Once a drug has been approved for one purpose, physicians can prescribe it for another. For instance, he points out, doctors now routinely prescribe aspirin to thin the blood and prevent heart attacks. Yet its approved use is as an anti-inflammatory drug for headaches, pain, and fever. It is much further afield, he points out, to use aspirin for heart disease than to use growth hormone for aging. "In one case you are talking growth hormone deficiency in children and in the other case, you are talking growth hormone deficiency in adults." The FDA accepted his argument, and in 1994, he says, he became the first doctor in the United States to openly offer growth hormone replacement.

More than 800 people have now passed through his clinic, including movie stars, high-powered executives, and a surprising number of physicians. According to Chein, there has not been one treatment failure. "It's a slam dunk," he says. "This is 100% effective. I tell my patients if I can't get the blood level of your hormones to look like that of a 20-year-old, you can get all your money refunded." He also guarantees results: a gain in bone density of 1.5 to 2.5 percent every six months as shown on bone scan; a loss of 10 to 12 percent of body fat every six months if they are overweight; and a gain of 8 to 10 percent in muscle mass every six months. The changes in body fat to lean mass ratio will continue, he says, until the body composition has reverted to that of a twenty-year-old. And then it will stay that way.

In 1995 Chein began collaborating with L. Cass Terry, M.D., Ph.D., chairman and professor of neurology at the Medical College of Wisconsin in Milwaukee. A neuroendocrinologist, Terry has conducted research in brain regulation of growth hormone. He was also a colleague and friend of Daniel Rudman, who was at the same institution, and began a pilot research project with Rudman on neurological disorders and IGF–1 that was left uncompleted as a result of Rudman's untimely death.

An athlete and bodybuilder, the fifty-five-year-old scientist and physician became interested in growth hormone replacement for himself when he learned that a fifty-five-year-old friend who was a former world-class weight lifter was going to Chein's clinic. "I'm fairly fit, lift weights, do a lot of aerobics but my body fat composition, like a lot of older males, had changed. I had gotten that little belt around the middle," Terry says. Unwilling to let nature take its course, he decided to go on growth hormone for himself when Chein told him that he could accomplish the same results as Rudman with no side effects.

In less than two months, Terry could feel the difference. "I lost a lot of the fat with growth hormone. I don't think my energy level has changed, because I've always been high energy. But my libido has changed significantly along with my sexual performance. I don't get nearly as winded or exhausted when I work out now. The dumb bell press on which I used to lift 30 pounds, I'm now up to 50 pounds and I can kill it."

He also noticed "little things," he says. "My nails and my hair grew a little faster and the skin around my eyes has a little more turgor from the water retention. My wife noticed that I have fewer wrinkles."

LOW-DOSE, HIGH-FREQUENCY GH INJECTIONS

Chein and Terry believe that the clinical results in their study (see below) are due to the method of HGH they have developed.

Their goal was to increase the blood concentrations of IGF–1, the hormone stimulated by the breakdown of growth hormone in the liver, which actually does most of the work of GH in the body (more about this in the next chapter). The level of IGF–1 they chose was 350 nanograms per milligram, the same amount that Rudman found "increased lean body mass, decreased body fat, increased vertebral bone density and increased skin thickness," say the researchers. But there were problems with Rudman's approach, which even Rudman admitted. Although it brought about the desired bodily changes, it caused serious side effects because it did not mimic the body's normal pattern of GH release by the pituitary.

Normally HGH is secreted daily in bursts, with the largest amount released during sleep. But Rudman gave the men in his study injections at a low frequency (three times a week) in high doses (approximately 16.5 international units [IU]—a standardized measurement of hormones, vitamins, enzymes and other biological substances. An IU of GH is 3 milligrams.) The identical high-dose, low-frequency regimen was adopted in a recent study by Maxine Papadakis and her colleagues of the University of California in San Francisco, which also showed a high incidence (77 percent) of side effects, such as lower-extremity edema and pains in both small and large joints. Interestingly, the research team found that the side effects "disappeared or remitted markedly within two weeks after the growth hormone dose was decreased by 25% to 50%"—the dosage level that Chein and Terry believe should have been used at the outset. There was no case of carpal tunnel syndrome in this study.

Chein and Terry have gone in exactly the opposite direction with their high-frequency, low-dose (HF-LD) strategy. They used one-quarter to one-half of Rudman's weekly dosage, 4 to 8 IU a week, or .3 to .7 IU twice daily. And instead of giving it three times a week, they taught their patients to administer HGH subcutaneously before sleep and upon arising six out of seven days a week. The seventh day without HGH is to ensure that the pituitary does not become "lazy" and stop releasing whatever growth hormone the body still produces. The nighttime and morning injections cause two daily bursts of GH, one during sleep and the other during the day. This more closely approximates the normal pattern of release and results in essentially no significant side effects, they found.

Chein asked Terry if he would use his skills as a researcher and academic to help him present the torrent of clinical data he had amassed from more than 800 patients who had been treated for up to two years, using the HF-LD technique of growth hormone administration. In this way, he hoped to convince both scientists and the public that treatment with human growth hormone could make a profound difference in reversing many of the degenerative effects of the aging process.

THE STUDY

The preliminary results of Terry's analysis drawn from more than 800 patients treated for up to two years between 1994 and 1996 are presented here for the first time in published form. "The HF-LD strategy effectively increases IGF–1 to youthful levels within 1 to 2 months by adjusting the weekly dose between 4 and 8 IU per week (recombinant biosynthetic HGH from Eli Lilly)," write Chein and Terry. The average somatomedin C (another name for IGF–1), blood levels from a sample of forty-seven patients are shown in the pre-

Effects on HF-LD HGH of IGF-1

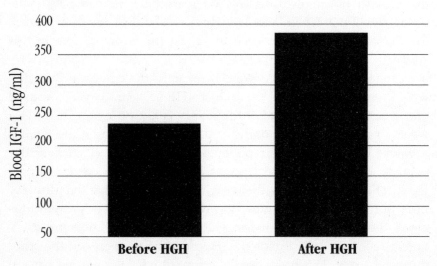

ceding chart and the accompanying table. The levels increased markedly by 61 percent. "These results demonstrate that HF-LD administration of HGH effectively increases somatomedin C to more youthful levels," say the authors.

SOMATOMEDIN C (IGF–1) TABLE

Effects of Human Growth Hormone Administration (High-Frequency, Low-Dose) on Somatomedin C Blood Levels

L. Cass Terry, M.D., Ph.D., and Edmund Chein, M.D., Medical College of Wisconsin and Palm Springs Life Extension Institute

Somatomedin C Blood Levels (ng/ml)

	Before HGH	*After HGH*	*Increase*[1]
Mean	238.8	384.5	61.0 percent
STDEV*	62.3	50.8	
* Standard deviation			
SEM#	9.1	7.4	
# Standard error of the mean (standard deviation divided by the square root of the sample size)			
No. Patients	47.0	47.0	
MAX	366.0	574.0	
MIN	132.0	301.0	

[1]$p<.001$

CLINICAL RESULTS

To determine the effects of the HF-LD method of growth hormone therapy, Chein and Terry initially analyzed 308 randomly selected questionnaires from 202 patients treated between 1994 and 1996. They ranged in age from thirty-nine to seventy-four years old, with women making up 15 percent of the study population. The exciting results of this study are shown in the table below labeled "Assessment." According to Chein and Terry, "The most outstanding results were improvements in muscle strength, exercise endurance, and loss of body fat. Also, there were significant improvements in skin, healing capacity, sexual drive and performance, energy level, emotions/attitude, and memory. In general, these improvements occurred within 1 to 3 months and continued to increase over 6 months."

ASSESSMENT

Effects of Human Growth Hormone Administration (High-Frequency, Low Dose) in 202 Patients[1]

L. Cass Terry, M.D., Ph.D., and Edmund Chein, M.D., Medical College of Wisconsin and Palm Springs Life Extension Institute

Strength, Exercise, and Body Fat	Improvement[2]
Muscle Strength	88 percent
Muscle Size	81 percent
Body Fat Loss	72 percent
Exercise Tolerance	81 percent

Skin and Hair	
Skin Texture	71 percent
Skin Thickness	68 percent
Skin Elasticity	71 percent
Wrinkle Disappearance	61 percent
New Hair Growth	38 percent

Healing, Flexibility, and Resistance	
Healing of Old Injuries	55 percent
Healing of Other Injuries	61 percent
Healing Capacity	71 percent
Back Flexibility	83 percent
Resistance to Common Illness	73 percent

Sexual Function	
Sexual Potency/Frequency	75 percent
Duration of Penile Erection	62 percent
Frequency of Nighttime Urination	57 percent
Hot Flashes	58 percent
Menstrual Cycle Regulation	39 percent

Energy, Emotions, and Memory	
Energy Level	84 percent
Emotional Stability	67 percent
Attitude Toward Life	78 percent
Memory	62 percent

[1] Mean self-assessment time was 180 days after HGH therapy initiation: range = 15–720 days.

[2] Based upon 308 responses: rating improvement as slight to definite.

Chein and Terry strongly believe as we do that HGH replacement therapy should be combined with other hormone replacement regimens as well as aerobic and resistance training, dietary evaluation, and mental fitness/stress reduction to achieve maximum benefit and deter the aging process (see the Growth Hormone Enhancement Program in the second half of this book). "Our patients are highly motivated people who are usually in the higher socioeconomic range," says Terry. "They do all the right things to keep their growth hormone at high levels. If I went to a Veterans' Hospital and took a sample of people who drink, smoke, and abuse their bodies and put them on HGH, they would not do as well."

SIDE EFFECTS

To date, Chein and Terry have not had any major side effects, including carpal tunnel syndrome, in the entire series of 800 patients. The only side effects that have occurred are minor joint aches and pains and fluid retention. These disappear in the first month or two. They believe that the persistent side effects, such as carpal tunnel syndrome, that occurred in other studies is due to overly high doses given in a nonphysiological manner.

GROWTH HORMONE DID NOT CAUSE CANCER

Most significant is the fact that there were no reported cases of cancer among all the patients treated at the clinic. This is particularly reassuring since some investigators have been concerned that growth hormone could cause undetected cancer cells to divide more rapidly. "With 800 people over the age of about 40," says Terry, "you would think that given the normal incidence rate of cancer, some of these people would get cancer. It could be that there is some sort of protective effect from growth hormone replacement."

Even more compelling, the levels of prostate specific antigen (PSA) levels, a marker of prostate problems including cancer, did not increase among any of the male patients. And in one case study, which Chein and Terry are writing up for scientific publication, growth hormone actually seems to have reversed the course of prostate cancer. The patient came to Chein with PSA levels in the 50 to 60 range (normal is 0.0 to 4.0 and men with cancer usually have PSA levels in the 10s or 20s). The diagnosis of adenocarcinoma of the prostate was confirmed on needle biopsy. Although existing cancer is normally a contraindication for hormonal replacement therapy, the patient refused surgery, and instead urged Chein to treat him with growth hormone along with DHEA and melatonin, but not testosterone. The man's PSA levels are now in the range of 5 to 7.

"It is mind-boggling," says Terry, who admits that he does not know why this apparent remission occurred. Chein speculates that the growth hormone stimulated the immune system, including the natural killer cells, which effectively destroyed the cancer cells.

WHAT HAPPENS WHEN YOU GO OFF GH?

At this point if your pituitary is no longer releasing growth hormone, there is no way to maintain high levels of GH and IGF–1 without injections of HGH. (This may change in the near future as oral secretagogues become available—see Chapter 21.) But because of the high cost of HGH, some people elect to have six months on and six months off the hormone, with some couples alternating in six-month intervals. With the kind of diet and exercise regimens outlined in this book, most patients are able to more or less maintain their changes in body composition, according to Terry. But they might lose their energy level. "I forgot to take HGH on a trip recently and I could tell the difference in my energy level," he says.

"I truly believe that growth hormone has made a huge difference in my life," says Terry. "There are people I know from Milwaukee who are on it and are all doing well. The big question is, is this *the* aging hormone? There are so many things, the loss of cells, the decrease in cell function. I think that growth hormone is a fantastic solution to a lot of aging problems particularly if you look at it from a socioeconomic point of view," he says. "If it keeps you healthier longer as you age, it is going to cut down on Medicare. I think it is a real revolution. But it has to be done right and shown scientifically."

Why is it that a change in a single hormone can have such incredible far-reaching effects throughout the mind and body? To answer those questions, let's next take a closer look at the biology of growth hormone and its role in aging and longevity.

The Biology of Growth Hormone

WHAT IS A HORMONE?

Have you ever wondered how the 100 trillion–plus cells of the human body function together as a unit? The tactical logistics of getting that vast army to integrate and coordinate their operations would be beyond the scope of all the greatest generals in history. Yet the body carries out millions of operations every second in its cells. It does this with the help of hormones (from the Greek *hormone*, "to set in motion"), tiny scurrying chemical messengers that move continuously through the bloodstream to different cells in the body. Once they reach their destination, the hormones bind onto areas on the surface of the target cell, known as receptors, where they stimulate a specific activity. Hormones are involved in every aspect of human activity from sexual function to reproduction, growth and development, metabolism and mood. A well-known example of a hormone is insulin, produced by endocrine cells in the pancreas, which digests food. Another is adrenaline, which stimulates the body during stress to deal with perceived threats.

The hormones are produced by the endocrine glands. The major endocrine glands are the pituitary, thyroid, thymus, adrenals, pancreas, ovaries, and testes. Located in the exact center of the brain, the pituitary gland is often called the master gland because it controls the release of many of the body's hormones. About the size of a kid-

Long-Life Hormones at a Glance

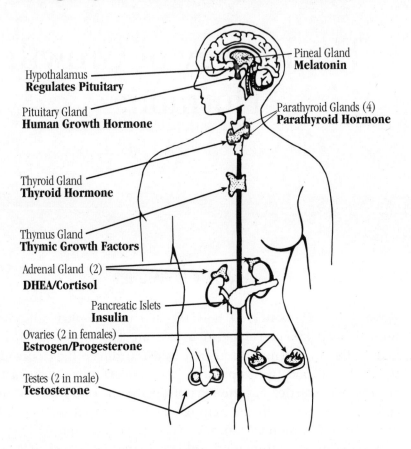

Pineal Gland
Melatonin

Hypothalamus
Regulates Pituitary

Pituitary Gland
Human Growth Hormone

Parathyroid Glands (4)
Parathyroid Hormone

Thyroid Gland
Thyroid Hormone

Thymus Gland
Thymic Growth Factors

Adrenal Gland (2)
DHEA/Cortisol

Pancreatic Islets
Insulin

Ovaries (2 in females)
Estrogen/Progesterone

Testes (2 in male)
Testosterone

ney bean, it has three lobes. The lobe that interests us in this book is the anterior lobe, which is the source of growth hormone. The anterior pituitary actually secretes ten hormones, which are required for the regulation of growth, reproduction, and metabolism.

THE HYPOTHALAMIC-PITUITARY AXIS

The pituitary is connected by a network of tiny red blood cells called capillaries to another brain structure called the hypothalamus. This is the command center in the brain. In terms of function, the pituitary with its ten hormones is like a "blind" instrument panel, indifferent to what is happening in the outside world. For that, it depends on the hypothalamus, which communicates with the outside by means of the signals sent to it by the nervous system. About the size of four peas together, the hypothalamus is a remarkable jack-of-all-trades. Among other things, it controls the stimulation of muscle fibers, the growth of the body via the secretion of growth hormone, thyroid gland function, the mammary glands, the sleep center, emotions, appetite, heat production, and the thermoregulation of the body.

The hypothalamic-pituitary axis regulates the activity of growth hormone, the sex glands, thyroid glands, and the adrenal glands, which, in turn, control the body's stress response. This control can be seen in the following diagram.

Hormonal regulation is a complicated feedback affair, in which one thing causes another, which causes another, in Rube Goldberg

The Major Endocrine Systems

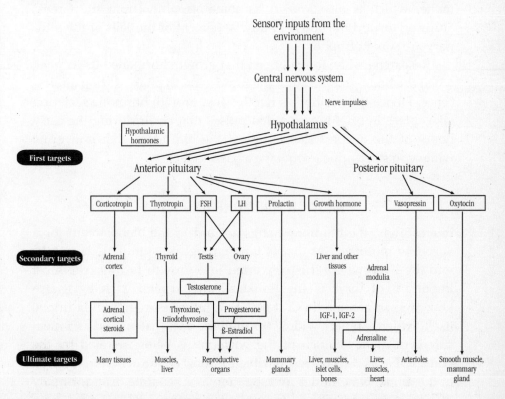

The major endocrine systems and their target tissues. Signals arising from the nervous system are passed via a series of relays to the ultimate target tissues. In addition to the systems shown, the thymus and pineal glands, as well as groups of cells in the gastrointestinal tract, also secrete hormones. FSH designates follicle-stimulating hormone; LH, luteinizing hormone.

fashion. The way it works is this. The hypothalamus secretes releasing factors, which in turn control the hormones released by the pituitary. The pituitary hormones then stimulate the secretion of hormones from the thyroid, adrenals, and gonads (the testes or the ovaries). When the concentrations of these hormones rise in the blood, a signal is sent to the hypothalamus to turn down the production of its releasing hormones, which then causes the pituitary to decrease its hormone production. The pituitary makes less of its hormones, which, in turn, slows the production of the hormones made by the thyroid, adrenals, and gonads.

GROWTH HORMONE SECRETION

The most abundant hormone made by the pituitary gland is growth hormone. Pituitary cells called somatotrophes make growth hormone, which is also known as somatotrophin (from the Greek, "turning toward the body.") Fully 50 percent of the cells of the pituitary are somatotrophes.

Researchers had long noticed that growth hormone hits its peak when the body goes through the rapid growth phase during adolescence. Hence the hormone's name. Most growth hormone secretion takes place in brief bursts called pulses that occur during the early hours of the deepest sleep. Indeed the old adage that you grow while you sleep appears to have a basis in fact.

GROWTH HORMONE AND IGF–1

Interestingly, the hormone hangs around in the bloodstream for a very few minutes. But that is long enough to stimulate its uptake into the liver, where it is converted into growth factors. The most important of these is Insulinlike Growth Factor 1 (IGF–1), also known as somatomedin C. It is IGF–1, rather than growth hormone itself, which can vary widely through the day, that is used as a measurement of how much of the hormone is being secreted by the body. IGF–1 is directly responsible for most of the actions described in this book. Now under investigation as a separate drug for many of the same indications of growth hormone, IGF–1 may even become the hormone of choice in the future (see Chapter 21).

There are two yin-yang factors that regulate the release of GH.

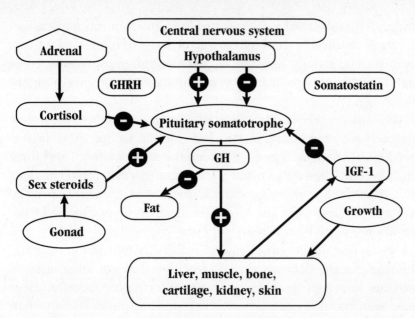

The central nervous system controls the hypothalamus, which in turn controls the pituitary somatotrophe (cells which secrete growth hormone) through two regulatory systems—GHRH (growth hormone–releasing hormone) which stimulates GH and somatostatin which blocks GH. Other hormones act as modulators. The adrenal glands make cortisone (the stress hormone), which inhibits GH, while the gonads (ovaries and testes) make sex steroids, which stimulate GH. The diagram also shows that GH acts on the liver to produce IGF-1, which in turn stimulates growth in a variety of tissues, including muscle, bone, cartilage, kidney, and skin. Finally IGF-1, acting through a negative feedback control mechanism, inhibits further production of GH. (Adapted from: Corpas, E., Harman, M.S., and Blackman, M.R. Human growth hormone and human aging, *Endocrine Reviews* [Feb. 1993] 14:20–39.)

One is growth hormone–releasing hormone (GHRH), which stimulates its release, and the other is somatostatin (SKIF), which inhibits its release. There are other factors that jack up GH release (see Chapter 17), as well as factors that block it. For instance, exercise, stress, emotional excitement, and dieting enhance GH release, while obesity and free fatty acids act as inhibitors. The complex control of growth hormone can be seen in the above illustration.

GROWTH HORMONE AND AGING

Growth hormone declines with age in every animal species that has been tested to date. In humans, the amount of growth hormone after the age of twenty-one to thirty-one falls about 14 percent per decade, so that the total twenty-four-hour growth hormone production rate is reduced by half by the age of sixty. In numerical values, we produce on a daily basis about 500 micrograms at twenty years of age, 200 micrograms at forty years, and 25 micrograms at eighty years of age.

As mentioned above, the easiest way to measure growth hormone in the body is by measuring plasma IGF–1 levels. Under 350 IU is considered evidence of deficiency. Between the ages of twenty and forty years, less than 5 percent of healthy men have less than 350 IU per liter of IGF–1 levels. But after age sixty, 30 percent of apparently healthy men have this low amount. And after age sixty-five, about half the population is partially or wholly deficient in growth hormone.

The decline in growth hormone is directly tied to the bulging, wrinkling, saggy, flabby, draggy creatures that we all sooner or later start to see in the mirror. Those of us with naturally lower amounts of the hormone age much faster and more visibly than those of us who by reason of genes or high exercise level maintain a higher level of secretion for a longer period of time. The loss of the hormone with age is similar to that seen with menopause. And it has been given a similar name, the somatopause.

Daily Human Growth Hormone Secretion

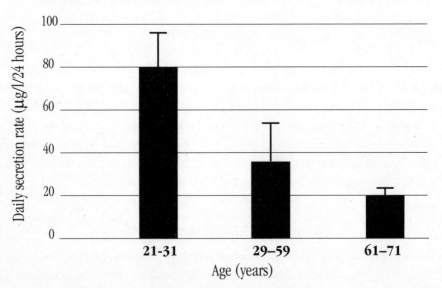

Daily HGH secretion rate is reduced progressively with increasing age.

Growth Hormone Decline

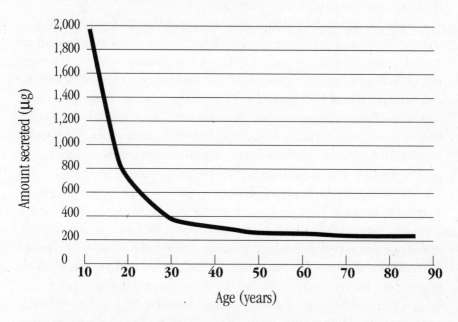

Total daily HGH secretion declines with age so that "elderly" levels are reached by age 35–40. Adapted from: the *Journal of NIH Research*, April, 1995.

Interestingly, studies of patients with pituitary disease show that the loss of hormones in that gland follows a particular sequence. First growth hormone goes, followed by the gonadal hormones, luteinizing hormone (LH), and follicle-stimulating hormone (FSH), and finally the thyroid-stimulating hormone (TSH) and adrenocorticotrophin (ACTH). So the loss of growth hormone from the pituitary is the start in a downward cascade of the pituitary hormones.

WHY DOES GH DECLINE WITH AGE?

The answer to this riddle has yet to be solved. Studies have shown that the aging pituitary somatotrophe cell is still able to release as much growth hormone as the young cell if it is adequately stimulated. This means that the fault must lie somewhere in the factors that regulate its release. Something happens in the feedback loop between the release of IGF–1 in the liver and the hypothalamus. Ordinarily a decline in IGF–1 tells the hypothalamus in the brain to direct the pituitary to make more growth hormone. But this nice feedback loop breaks down with age.

Some researchers believe that the problem lies with somatostatin, the natural inhibitor of growth hormone. It has been found to increase with age and may act to block the secretion of growth hormone. When researchers knocked out the action of somatostatin in old rats, they had GH pulses that were as large as those of young rats. Other researchers believe that the precursor hormone, growth hormone–releasing hormone (GHRH), which stimulates the release of growth hormone, becomes less responsive to feedback signals. It is also possible that both things may be happening.

The effect in the body is similar to what happens with another hormone, insulin. With age, we become less sensitive to insulin. As a result, we do not metabolize glucose as efficiently, and consequently there is a rise in blood glucose. In about a third of the older population, this insulin resistance, which is related to the kind of spare-tire obesity seen with aging, is severe enough to be a disease—type 2 diabetes. Unlike the more familiar type 1 diabetes, the problem is not that the body doesn't manufacture insulin, but rather that the body's tissues act as though the insulin were not present. The latest thinking is that a similar phenomenon happens with growth hormone. Not only does the amount of growth hormone available to the tissues decline with age, but our tissues fail to respond to the growth hormone that *is* there. In this view, aging can be seen as a disease of growth hormone resistance in the same way that type 2 diabetes is a disease of insulin resistance.

THE DECLINE OF GH IS NOT INEVITABLE

The most recent research shows that whatever causes the decline of growth hormone, it is not irreparable and it is not permanent. William Sonntag, professor of physiology, and his colleagues at Bowman Gray School of Medicine in Winston-Salem, North Carolina, have recently completed an experiment that clearly shows the decline in growth hormone secretion with age is reversible. Old rats like old people have a decline in the bursts of growth hormone that are secreted. But when Sonntag and his colleagues took old rats at the age of twenty-six months and restricted their caloric intake, after two months the growth hormone secretion came back. In Sonntag's words, "We got a restoration of high amplitude growth hormone secretion, which indicates to us that it is not an anatomical problem or a permanent change within the hypothalamus that has happened." A few years ago, he showed that L-dopa, which is known to stimulate growth hormone, also restored the growth hormone pulses to their youthful level.

The take-home message is that the decline of growth hormone with age can be reversed. Even if the activity of growth hormone–releasing hormone declines, or the somatostatin increases, or the receptors become less responsive to the effect of growth hormone, these can all be overcome by the administration of growth hormone or growth hormone–releasing stimulants. In fact, as we will discuss later in Chapter 21 on secretagogues, recently developed manmade chemicals do the job even better than nature's own growth hormone–releasing hormone.

THE GLOBAL REACH OF GH

Growth hormone exerts its actions either directly or indirectly through its intermediary insulin growth factors to every organ system of the body. As we will see in the sections to come, almost nothing escapes its magic touch. In the same ways that it grows the bones of young children, it increases the size of most organs and tissue. Even the brain is affected. The latest studies in animals show that it can regenerate damaged brain tissue (see Chapter 12).

An abstract of a recent report by a group of researchers headed by Jens Sandahl Christiansen of Aarhus Kommunehospital in Denmark, sums up just how ubiquitous the hormone's effects are. "Untreated GH-deficient adults have been shown to have increased cardiovascular mortality, reduced exercise capacity, reduced muscle strength, subnormal glomerular filtration rate and renal plasma flow, defective sweat secretion and defective thermoregulation, reduced energy expenditure and basal metabolic rate, abnormal thyroid hormone metabolism, reduced myocardial function and clinical signs of premature atherosclerosis. Body composition has been found abnormal with increased fat mass, decreased lean body mass, decreased muscle fat ratio, visceral obesity, reduced extracellular fluid volume, and reduced bone mineral content. Furthermore, two independent groups have reported impaired psychological well being as compared to normal subjects."

The effect of growth hormone is not as dramatic as that of some of the other hormones in the body. A sudden drop in insulin levels, for example, can send you into insulin shock, which can be fatal. But a drop in growth hormone levels after age thirty or so can send the body into the slow decline, which, untreated, we call aging. Why do we age, and can growth hormone extend the quantity as well as the quality of life? These are the questions that we will look at next.

5

A Disease Called Aging

A few years ago the normally staid scientists at the Gerontological Society of America meeting in Washington, D.C., were in an uproar. Angelo Turturro, Ph.D., a senior scientist at the National Center for Toxicological Research in Jefferson, Arkansas, made the stunning announcement that everything we called "aging" was no more than a collection of diseases and pathologies. That went for everything from a rise in blood glucose and blood pressure to cataracts and skin wrinkling. Knock out each one of these disease processes and there was no telling what life spans human beings could reach.

People began jumping to their feet, demanding to be heard. One indignant young woman yelled out, "What about menopause? Is that a disease, too?" Yes, Turturro declared, there were pathological changes in the hormonal balance that terminated a woman's fertile period. Fix this, and motherhood could occur at any age.

Extreme as his position sounded, he was basing it on experiments that he and Ronald Hart, Ph.D., have done in their laboratory and similar studies all over the world. Rats and mice that have had their food intake severely restricted can live a very long time, even doubling their average life span. And these animals are amazingly free of diabetes, heart disease, cancer—all the plagues associated with aging. When they do get these diseases, it is much delayed when compared with animals who eat all they want. This research has recently been extended to monkeys. Even though it is far too soon to tell whether it will extend their lives, these animals are also showing the same favorable changes in such things as their blood glucose and blood pressure as the rodents do.

There is also evidence from human beings to support Turturro's point of view. Jeanne Calment, who rode a bicycle until age one hundred and smoked until age 118 and is still alive at 121, appears to have gone unscathed by major illness. In fact, according to Thomas Perls, M.D., of Harvard Medical School, who has been studying centenarians, this is true of most of the truly oldest old. "If you can live through the vulnerable septuagenarian and octogenarian years without Alzheimer's disease, a stroke or heart attack or other devastating illnesses, then you have gotten over the proverbial hump, and you can look forward to another 10 or 20 years of good health."

BIOLOGY IS NOT DESTINY

My own point of view is that there is a biologically controlled aging process, and it is designed to get Madame Calment and all of us in the end. If there were not such a thing as biologically controlled aging, then people who live on pure organic food, or reside on serene mountaintops, or chant mantras with the sunrise would be immortal. With their low stress, high level of physical movement, and good diet, they shouldn't die of anything. But, of course, that is not the case. They all age, because aging is biologically based.

There is an abundance of evidence for this. The cells of the body age just as the body does. When you put cells into a test tube and grow them, they divide only a finite period of time before they stop dividing. This is known as a Hayflick limit, named for the scientist, Leonard Hayflick, Ph.D., who discovered it. Once the cell stops dividing, it ages and dies. The ability of the cells to reproduce is closely related to life span. Long-lived species like humans have cells with a far greater division potential than those of short-lived critters like mice. Aging is also related to division potential. Old people's cells in culture putter out much faster than those of younger people, while cells from a fetus multiply like bunnies on a rampage until they too eventually run out of steam.

The latest evidence indicates that there is a "clock" or "counter" in each cell that is governed by a piece of DNA known as the telomere at the end of each chromosome in the nucleus of each cell. After each cell division, the telomere becomes a tiny bit shorter. When a critical amount of the telomere has been removed, the cell can no longer divide. Its metabolism slows down, it ages, and dies. New exciting research shows that the counter can be "turned on" and "turned off" in the cell. The control button appears to be an enzyme called telomerase which can relengthen the telomere and allow the cell to divide endlessly. Most cells of the body contain telomerase but it is in the "off" position so that the cell is mortal and eventually dies. But other cells are immortal because the telomerase is switched on. For instance, the hemopoietic cells, the progenitors of blood cells, are immortal. Unfortunately for us, so is cancer. Cancer cells do not age, according to the telomere theory, because they produce telomerase, which keeps the telomere intact. In the not-too-distant future, we will have telomere therapies that control the clock in each cell, blocking telomerase in cancer cells so that they stop dividing and die off and relengthening the telomere in aging cells so that they become young again.

There are other signs of aging in the cell. The proteins become gummed up in a process known as cross-linking. One form of cross-linking occurs when sugar molecules become attached to proteins and DNA, a process known as glycosolation, which can result in the aptly named accumulated glycosolation endproducts (AGE). Cross-linked proteins can cause cataracts in the lens of the eye, clog blood vessels, and jam the filtering mechanism in the kidneys. Free radicals generated by the breakdown of oxygen in the body damages cells, particularly the powerhouses of the cell known as mitochondria. The genetic material of the cell, the DNA itself, becomes torn apart and damaged with time and even though it is repaired, the repair processes can't keep up with the rate of damage.

But there is one way in which aging is exactly like a disease. Until very recently in human history, disease was thought of as a natural process. There was nothing you could do about it. Tuberculosis, smallpox, dysentery, the plague were just part of the natural order of things, God's will. All you could do was go to church and pray that it didn't happen to you. It wasn't until doctors understood that diseases were caused by something—viruses, bacteria, the environment—that it became possible to prevent or cure them. Now we are at the same point in regard to aging. Aging, which has long been thought of as inevitable, part of the human condition, can now be seen as a disease for which there are causes and treatments.

And like TB, smallpox, dysentery, and the plague, aging too is yielding to scientific discovery and medical intervention.

Fortunately we human beings don't accept the natural order very easily. If we did we'd still be living in caves and hunting with spears and stones. But we are an animal that changes its environment, and by changing our environment, we change nature. We have rewritten the rules of disease, and with growth hormone replacement, we are rewriting the rules of biology *and* aging.

Why am I so certain of this? In a May 1996 issue of the distinguished journal *Science*, researchers reported that they genetically engineered three genes associated with aging in the lowly worm known as the nematode, bumping up the life span of this species *fivefold*! The mutant nematodes looked and behaved normally—for a worm, that is—but they ate less, defecated less, and wiggled more slowly when they swam. The fact that they ate less seems to tie in with the idea that reduced food intake boosts longevity. But the big news is that a minor genetic alteration could achieve the equivalent of 375 years in human terms—proof positive that the life span is not etched in stone as conventional medical science has so long maintained.

DR. DILMAN'S HYPOTHESIS

The late Vladimir Dilman, an eminent Russian researcher in aging who spent his last years in the United States, proposed that the hypothalamic-pituitary axis formed a neuroendocrine "clock" of aging. In the same way that the telomere "clock" may control how many times a cell divides, the neuroendocrine clock times the rate of aging in the body's organ systems.

When we are young, he pointed out, the feedback system between the hypothalamus, the pituitary gland, and the other endocrine glands is under tight control. It is like a room thermostat. If the door is opened and the room cools down, the sensor in the thermostat picks up the change in temperature and the heat turns back on. But, with aging, this exquisite control slips and the endocrine system begins to respond less and less well to the needs of the body. The wonderful fine-tuning, continually adjusting mechanism known as homeostasis gets progressively out of kilter. The cells begin to function less and less well, and the result is the diseases of aging and eventually death.

Levels of Neuro-Endocrine Regulation

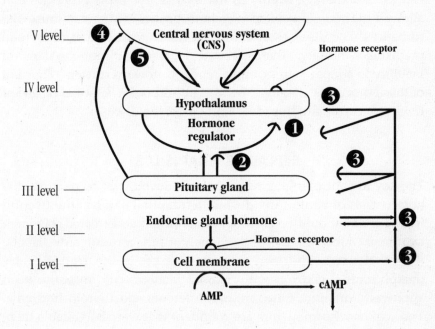

Levels of neuro-endocrine regulation in an organism:

1—intracellular level (scheme of cell membrane, hormone receptor (HR), and cyclic AMP (cAMP)-transmitter of action of hormonal signal;

II—level of peripheral endocrine glands;

III—pituitary level;

IV—hypothalamic level;

V—level of central nervous system; 1) ultrashort loop of feedback mechanism, effect of pituitary hormone on hypothalamus; 2) short loop of feedback mechanism, effect hypothalamic hormones on hypothalamus; 3) long loop of feedback mechanism, effect of hormones of peripheral endocrine glands and products of metabolism (glucose, fatty acids, etc.) on the pituitary gland and hypothalamus; 4) regulation of activity of central nervous system by the pituitary gland; 5) regulation of activity of central nervous system by hypothalamus. Adapted from: Dilman, V.M. *The Grand Biological Clock*, Mir Publishers, Moscow, 1989.

WHY DOES THE BRAIN CLOCK RUN DOWN?

It was Dilman's original contribution that the growth and development of the body contain the seeds of its own destruction. If the body were to stay in perfect homeostasis, he argued, everything would stay exactly as it was. Muscles would not expand, bones wouldn't grow. The problem is that after the period of development is over, at about age twenty to twenty-five, the same processes that allowed the body to deviate from homeostasis continue to cause disturbances. The diseases of disturbed homeostasis, he said, include prediabetes, obesity, atherosclerosis, depressed immune system, an inability to adapt to stress, and a tendency toward cancer. "The sum of these processes is aging," he said in his book *The Grand Biological Clock.* "According to this idea, aging itself is a disease."

REPLACING WHAT IS LOST

The key to controlling aging, Dilman believed, was to re-create the homeostasis of youth. Any deviation from that of a healthy twenty-five-year-old should be considered abnormal and treated. His own treatments included the use of phenformin, an oral anti-diabetic drug that was withdrawn from the market in the United States (metphormin, a drug with similar action, has recently been approved); dilantin; estrogen; progesterone; and thyroid hormone. Those of us who knew him are confident that were it available then, he would have included growth hormone in his regimen.

Dilman's work inspired many anti-aging physicians, including me. He was one of the first people to view aging as a treatable disease. His ideas, which were first published in the 1950s, set the stage for the kind of intervention that will end aging as we know it.

DR. WONG'S HYPOTHESIS

Growth hormone may actually stop aging on the most fundamental level of the body's organization—the cell, according to Grace Wong, Ph.D., a scientist in the department of molecular oncology at Genentech, the original makers of recombinant growth hormone. According to Wong, a great deal of aging is due to the breakdown of proteins in the cell as well as the DNA and RNA that provide the blueprint for making protein. Skin, hair, bone, and muscle are all made of proteins, and as these proteins degrade, so do the things that are made of them. The skin wrinkles, hair falls out, muscles shrink, bones lose their density. In the brain, the situation is even worse because the primary brain cells, the neurons, cannot regenerate. As a result, starting in middle age, our mental power declines.

Reaction time slows, short-term memory develops gaps, we can't race those *Jeopardy!* contestants to the questions the way we used to.

What makes the proteins of the cell degrade with age? A major cause, says Wong, is the generation of oxygen free-radicals in all cells that use oxygen to produce energy. When the oxygen is broken down in the cell, it can result unavoidably in the production of these short-lived, destructive particles. Wong proposes that these oxygen radicals, in turn, activate destructive enzymes called proteases that damage and degrade the proteins in the cell. When enough proteins are damaged, the cell dies, followed by fragmentation of the DNA.

Antioxidants, such as vitamin C and vitamin E, can remove the oxygen free-radicals and keep the proteases from becoming active. But growth hormone can act directly on the proteases by activating a cellular defense force called protease inhibitors. This means that even though oxygen free-radicals are still present in the cell, the protease inhibitors prevent them from doing their deadly work. Protease inhibitors, such as ritonavir and saquinavir, are now being used in the treatment of AIDS, and preliminary reports indicate that they bring the virus down to virtually undetectable levels in affected people. (See Chapter 8 on the use of GH in AIDS.)

In laboratory experiments, growth hormone was able to protect animals against the free-radical–killing effects of radiation and hyperoxia. In one experiment, the animals were given a mixture of 98 percent oxygen, which at this concentration is ordinarily toxic over time and quite lethal. But when they were given growth hormone, they all survived! Wong believes that the decline in growth hormone with age may be a major factor in the loss of proteins that occurs in later life. "As you age," she says, "you are releasing less growth hormone, your immune response is decreasing, and at the same time the amount of oxygen free radicals are increasing." Without growth hormone in the cell to induce protease inhibitors, the proteases activated by the oxygen free-radicals are free to tear up the cell's proteins.

Growth hormone may serve yet another important anti-aging function, she believes. It may actually stop *apoptosis*, programmed cell death (more about this phenomenon in Chapters 8 and 12). This cell suicide is brought on not by random events like the generation of oxygen free-radicals, but by the cell's inner clock telling it that it is time to die. A great deal of apoptosis takes place during aging and fetal development; indeed the development of an organism from a single fertilized egg could not occur without the large-scale death of cells as the organs are formed from masses of cells, the way a sculptor chisels away at a mass of marble. The same process is believed by many scientists to occur in later life, resulting in the programmed cell death in the heart, brain, bone marrow, and elsewhere. But in this case, instead of the emergence of new life, loss of function occurs, with death the final result. Apoptosis robs us of our memories, our energy, our zest for life.

Here too, Wong believes that growth hormone has a role to play. Scientists have shown that when they introduce an apoptosis-associated protease gene into brain cells, the cells die. But when they cause cells to make protease inhibitors that block apoptosis, the cells survive. By stimulating the cells to make protease inhibitors, growth hormone can stop apoptosis, or programmed cell death.

The bottom line is that growth hormone stops aging in the cells in two ways: First, it stops the random wear and tear on the cell

Growth Hormone Promotes Cell Survival

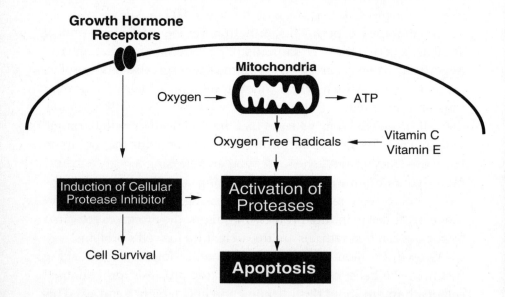

Mitochondria (the powerhouse of the cell) use oxygen to produce ATP for cellular energy. Free radicals formed by the breakdown of oxygen can activate protein-destroying enzymes called proteases, which can cause apoptosis (cell suicide). Antioxidants, such as Vitamin C and Vitamin E, can help the cell by scavenging the oxygen free-radicals. Growth hormone can work to generate protease inhibitors that may block the proteases and allow the cell to survive. *(Diagram courtesy of Grace Wong, M.D.)*

caused by oxygen free-radicals, and second, it arrests apoptosis, the cells' built-in self-destruct program. If Wong's hypothesis is right, no wonder growth hormone is such a powerful anti-aging tool!

The latest research—some of it so new that it is yet to be published—shows that growth hormone not only affects the architecture of the cell, but works on the blueprint of the cell itself—the DNA.

GETTING AT THE BLUEPRINT OF DNA AND AGING

"The blueprint of aging is in the DNA under the hood of the telomere,"—the "clock" at the end of every chromosome that is shortened with each cell division, says noted plastic surgeon and antiaging researcher, Vincent Giampapa, M.D., director of clinical research at the Longevity Institute International in Montclair, New Jersey. To actually reverse aging at the cellular level, we will need a substance that will restore telomere length and like a genie turn old cells into young ones. That is not yet available, although Giampapa believes it will be in less than a decade. Until then, growth hormone and its attendant hormone, IGF-1, can do the next best thing—help keep the cell in as healthy a state as possible.

The cells' ability to function depends on the genetic material, the DNA, in the nucleus of the cell which codes for all the proteins, hormones, and enzymes that make the cell run. The DNA is like an army under constant attack from oxygen free-radicals, ultraviolet light, the heat of the body, and other damaging factors. Although the DNA has the ability to repair itself, it falls down on the job with age, a victim of the same aging process that affects the cell. At the same time, damage is accumulating in the energy center of the cell, the mitochondria, which has its own DNA. Up to now, one of the few ways we could limit the damage to the DNA was to take antioxidant supplements such as vitamin C and E to bolster our own defenses.

But, according to Thierry Hertoghe and Giampapa, the latest European research shows that growth hormone and IGF–1 can go further and do what antioxidants cannot. Growth hormone and IGF-1 act like carriers to bring the cell the raw materials needed for renovation and repair. IGF-1 launches the delivery of the nucleic acids, DNA and RNA, right into the cell nucleus, where the DNA resides. The nucleic acids are used to repair damage to the DNA and stimulate cell division. Growth hormone initiates the transport of amino

acids, the building blocks of protein, and nucleic acids into the cyto-
plasm of the cell, the area outside the nucleus. This includes the cell
membranes and intracellular organelles, such as the mitochondria. In
this way, growth hormone and IGF-1 don't just minimize the damage
to the DNA and cellular structures, they help heal the cell and the
DNA. These two hormones actually treat the blueprint of aging.

GROWTH HORMONE AND LONGEVITY

Does growth hormone extend the life span? When gerontologists
speak about life span, they are really talking about two different
kinds. One is the *average life span*, the age attained by 50 percent of
the population. This has tripled since Roman times and is now
about seventy-seven years in the United States. The other is the
maximum life span, the age attained by the oldest individuals in a
population. Right now, this appears to be about 121 years old. Most
gerontologists would like to "rectangularize the curve," that is,
move the median life span closer to the maximum so that at least 50
percent reach the post–one hundred mark.

    This is what Blackman, who is running the growth hormone tri-
als at Johns Hopkins, believes will happen with interventions, such
as GH. "We are adding years to life towards the intrinsic life span of
people. We don't believe that any intervention is going to change
the intrinsic maximum life span of people, but rather push people

Life Expectancies in the U.S.

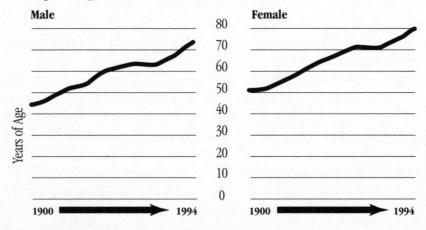

in a healthier fashion towards that maximum life span, which is somewhere between 110 and 120 years."

As we will see in the accounts of the men and women who are taking HGH, a kind of life extension is already occurring. If you go by how you feel when you wake up in the morning, awake and eager as a child ready to start the day, rather than feeling like you have to drag yourself through another twenty-four hours, then in terms of what you can accomplish, using growth hormone is like being handed twenty years. Just think, how many times have you said to yourself, if only I were thirty years old, I'd start a new business; build a house; write a novel; sail around the world; take up scuba diving, horseback riding, or in-line skating. With growth hormone replacement therapy, there is no longer any reason to feel that you are too old to do anything you have dreamed of doing. You can have all those years back again.

There is abundant evidence, as you will see in the succeeding chapters, that growth hormone therapy decreases your chances of dying prematurely from disease, which is another way of saying that it increases your chances of celebrating your 120th birthday. In animals, it has been shown to regrow the most important organ of immunity—the thymus (see Chapter 7). If growth hormone can restore the thymus in human beings, and there is no reason to believe that it won't, then we should enjoy the same freedom from illness in old age that we do at age ten—the highest point of immune function! And as detailed in Chapter 8, it also combats disease both by lowering risk factors and by reversing their course if you are already sick.

How can we be so confident that growth hormone treatment will extend the average life span? Because we know what hormonal replacement can do. The results of the largest life extension experiment in history are already coming in. Estrogen replacement therapy in postmenopausal women has cut their rate of heart attack and stroke in half and prolonged the life span. (See Chapter 15.) Growth hormone will perform even better, since its reach is far more pervasive in the body.

Thierry Hertoghe, M.D., a Belgian physician working in the field of endocrinology, has no doubt that growth hormone has a profound effect on human aging and longevity. "When you look at Bengtsson's study of 333 people with pituitary insufficiency," he says, "they died at twice the rate of normal people. [See Chapter 2.] An adult at the age of 50 has the same production of growth hormone as a young adult who is GH-deficient. So the 50-year-old has twice as much chance to die compared to a person who is treated with growth hormone. I find it sad that people are aging, have a lesser quality of life, and die prematurely when they could be on hormonal treatment. This is the medicine of the future."

EXTENDING MAXIMUM LIFE SPAN

But the big question is whether growth hormone will extend the outer boundaries of human survival. Many gerontologists, myself included, contend that the maximum life span is not immutable. With advancing technology, such as the ability to engineer the genes at will, we will push back the frontier to 150 years and beyond. Interventions like growth hormone will allow us to live long enough so that the new interventions that will achieve the next leap in life span will be on board. But although the evidence is very preliminary, there are some strong indicators that growth hormone may not only extend the *quality* of life but the *quantity* as well.

At present, the only practical way to test a life span intervention is to use a short-lived animal species, preferably a mammal, that bears some resemblance to the human situation. In 1990, two researchers at North Dakota State University, David Khansari, Ph.D., and Thomas Gustad, Ph.D., attempted to answer this question. They gave growth hormone injections to a group of twenty-six mice that were seventeen months old, or more than three-quarters through their average life span of twenty-one months. The animals were showing signs of aging and members of the original colony of sixty mice had started to die off. A control group of twenty-six mice who were the same age received placebo injections of saline solution. After thirteen weeks, sixteen animals, or 61 percent in the control group had died, while all but two, or 97 percent, of the growth-hormone treated animals were still alive! In other words, the vast majority of the treated animals had already lived longer than the life expectancy for that species.

At this point in the experiment, the researchers sacrificed four animals from each group to look at their immune function. The remaining mice were kept alive untreated for another four weeks. During this time, the control group had completely died out, while only one mouse from the hormone-treated group died. Khansari and Gustad resumed GH therapy for another six weeks until they had exhausted their supply of growth hormone and were forced to end the experiment, sacrificing all the animals. Only one mouse died during this period of the study. This means that out of the original group of 26 treated mice, only four mice died of natural causes. This left 18 mice (4 having been sacrificed) who were still alive 22 weeks after treatment had begun, while the entire control group (less four animals that were sacrificed) had perished entirely after 16 weeks (see figure below). The results, say the researchers, "suggest that long-term GH therapy prolongs the average life expectancy of the hormone treated mice significantly."

The treated animals not only looked younger, they had more youthful immune systems as measureed by several stardard tests (see Chapter 7). The researchers had begun their study when the animals were already showing signs of aging and deterioration of their immune system, including cancer and infection. "Thus," they write, "the observed prolongation of the mean life expectancy in the

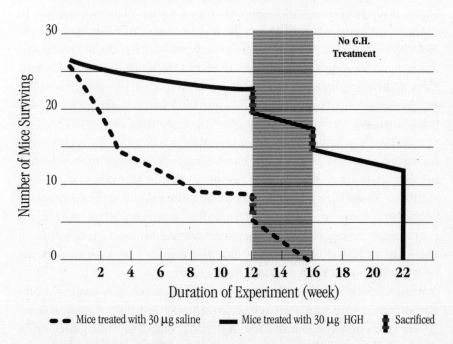

Mortality Curve of 19-Month-Old Mice

No G.H. Treatment

Number of Mice Surviving

Duration of Experiment (week)

● ● Mice treated with 30 μg saline   ━━ Mice treated with 30 μg HGH   ▌ Sacrificed

Both treated twice per week for the period shown. Adapted from: Khansari, O.N. and T. Gustada, *Mechanisms of Aging and Development* 57 (1991): 87–100.

hormone-treated group seems to be due to a delay or prevention of age-associated disorders."

One can only wonder what would have happened had the researchers been able to carry out their experiment until all had the mice had died. Would they have been able to extend the *maximum life span* for that species? There are very few treatments that have ever been able to accomplish that feat. And the very best method, which has given consistent results across the animal kingdom from single celled protozoans to fruit flies to rodents is reduction of food intake.

In well carried out experiments, animals on a calorically restricted diet have a enjoyed a maximum life span that was more than twice the average life span for that species. This would be like a human being living beyond age 150!

Could a significant factor in these animals' ability to defy death be growth hormone? Dr. William Sonntag at the Bowman Gray School of Medicine at Wake Forest University in Winston-Salem, North Carolina, has examined what happens to growth hormone and IGF–1 secretion in animals that are diet restricted. Normally when we age, the amount of growth hormone and IGF–1 decreases along with protein synthesis—the manufacture of new proteins to carry out all the work of the cells and tissues. But Sonntag and his associates found just the opposite happened in the diet-restricted animals. Young rats on a moderate food-restricted diet actually had their growth hormone secretion go down, but by the time they reached twenty-six months—old age for a rat—their growth hormone pulses were equal to that of a young control rat.

"What we have tried to relate that to is that the calorically restricted animals at that age have a higher capacity to synthesize protein in their tissues," says Sonntag. While the rates of protein synthesis went down in old control rats, the aged restricted rats had 70 percent increase of new protein in the heart and 30 percent in the diaphragm compared with the unrestricted animals. Interestingly, the level of IGF–1 did not rise, but the number of receptors in the cells for IGF–1 increased by 60–100 percent.

Some people who are interested in life extension have already started their own food restriction programs, cutting back calories by 20–30 percent over what they normally eat. But this experiment shows that one of the important factors in life extension may be the increase in growth hormone. And as we will show in the second half of the book, you can accomplish the same goal without undertaking a spartan diet that may be next to impossible to maintain.

RAISING THE LIFE SPAN ROOF

At this time, the best therapy against aging is to limit the damage to the DNA with antioxidants, vitamins, and minerals, and to treat the DNA with the Growth Hormone Enhancement Program of diet, exercise, and GH releasers that forms the second half of this book.

Within five to fifteen years, we will have agents that will repair damage to cells, stimulate the telomere to regrow itself, and manipulate the DNA. At that point, we will have true age reversal. Old cells will become new, vibrant, active cells with the same division potential as young cells. Some scientists believe that research into telomeres will make it possible to reset the cellular clocks. With these advances, we will change the blueprint of aging, building a stronger, more durable edifice on the ruins of the cellular debris. It will be the end of aging and the beginning of a new era of limitless health and life span.

But the train is leaving the station, and anyone who is past age thirty-five or so will have to clamor on now. Listen to what Giampapa says: "If you don't limit cellular damage and treat the DNA now, in the next five to ten years when telomere stimulants and other DNA modifiers are on board, you will have so much extra damage to your chromosomes and DNA that you won't be able to benefit. The best prescription for people who are now in their 40s to 60s is utilize the best antiaging knowledge to minimize the damage to DNA so that they will still be able to benefit from antiaging therapies on the horizon. This is the bridging formula that will get you to the future."

And that future is golden indeed. We predict that the use of antioxidants, the first technology of anti-aging medicine, will add ten years to the average life span of seventy-seven years. Growth hormone and other hormone replacement therapies are the second tier of anti-aging technology that may add another thirty years to the life span. In the final chapter of this book, we will tell you how the coming changes in technology that are already on the drawing board today will extend the maximum survival of people who are around in thirty years to incredible potential life spans of 150 to 200 years or more! If that's not a good incentive to remain healthy and alive until 2025, I'd like to know what is.

6

The Global Impact of Growth Hormone: Case Studies

Most medicines are designed to achieve a specific end. Diet pills can make you lose fat, antidepressants can lift your mood, and sleeping pills can fight insomnia. But growth hormone can do each one of these things and so much more. The awesome power of GH to bring about total rejuvenation unlike any medicine that has ever existed can perhaps best be appreciated by the stories of the people who have experienced it firsthand. The men and women depicted in this chapter are all truly adventuresome souls. They are dreamers and risk-takers as well as movers and shakers. They are physicians, nurses, scientists, entrepreneurs, former models and bodybuilders, all of them linked by a single drive: to stop aging in its tracks.

HOWARD TURNEY (AGE SIXTY-FOUR)

We met Howard Turney, a businessman and entrepreneur, at the first meeting of the American Academy of Anti-Aging Medicine in Cancun in 1993. Turney had set up the first longevity clinic to dispense HGH for adults in North America in the town of Playa del Carmen, about ten miles outside Cancun. He named it, appropriately, El Dorado. Three years earlier, when he was fifty-nine and a half, Howard Turney, a six-foot, two-inch Texan with a personality as big as the Lone Star State, felt he was in terrible shape. He could barely tighten his belt around his forty-two-inch waist, he had palsy in his right hand, his skin was so thin "you could pull it up and it would just stay there in a heap," and he was a washout in the bedroom.

Then he read the article in the *New England Journal of Medicine* by Daniel Rudman on the treatment of twelve elderly men with growth hormone, and it changed his life. Since the FDA restricted the use of GH to pediatricians for the use of promoting growth in dwarfed children, Turney crossed the border into Mexico where the hormone was sold freely. His physician agreed to monitor the effects. He felt better almost immediately but he was afraid to attribute his response to anything more than a placebo effect, he says, because "I wanted it to work so badly." Then about three and half months into treatment, he had a dental disaster: four teeth pulled, three root canals, six hours of major oral surgery, with a temporary bridge put in.

Eighteen days later he was back in the dentist's chair because the bridge was loose on one side. He opened his mouth and the dentist peered in. "My God, Turney," he said. "What are you doing? You've healed more than two months in eighteen days."

"At that point," says Turney, "I knew that growth hormone was working. It regenerates cell growth in every part of your body. Osteoporosis is bone mass disease, so that had to be bone regeneration."

Turney can't wait to show off what the hormone has done for him. He proudly displays his trim thirty-three-inch waistline. "I haven't changed my diet or exercise program at all," he says. He rolls up his sleeve. "Feel that," he says, displaying a baseball-sized biceps of solid steel. He pinches the skin in the cushion between the base of the thumb and the hand. It bounces right back. A month before his two-year anniversary on the hormone, he absentmindedly began reading some small type without putting on his bifocals. To his amazement, he had no trouble. He picks up a matchbook now and reads the minuscule letters at the bottom, "Dimension National Company, Mass. 01041."

Then he holds up his right hand. "Before I went on the hormone, I had developed a palsy in this hand. It's the one I write with and shake hands with. It didn't happen all the time but when it did it drove me crazy. I was aggravated as hell. Four months into the treatment, it went away completely. My doctor had never heard of anything like it." His hand remains rock-steady throughout this speech.

But Turney isn't through. "My sex life is fantastic. I am as good as I was at twenty-five. At fifty-nine, I might go every four days if I was really loving somebody. And it was never as hard or satisfying as it used to be. Now it is better than it ever was," he says, grinning broadly at his girlfriend, who is about twenty years younger.

Paulene McNair, R.N. (Age Fifty-three)

She was the poster girl of growth hormone. At age fifty-one, the blond former model was shown on the cover of *Health* magazine rising from the water in Cancun like some latter-day Venus. A cardiovascular nurse specialist in Houston, Texas, she had been an operating room nurse for the heart transplant pioneer surgeon Michael DeBakey. In 1989 McNair became interested in rejuvenation, when she accompanied her father to the Ana Aslan Institute in Rumania to check out Aslavital (Gerovital). But it was a problem with her eye rather than aging that got her into growth hormone.

She had developed a huge gray hole in the vision of her right eye. All that was left was a tiny bit of peripheral vision in that eye, she says. The diagnosis was a form of nerve inflammation and the only treatment was cortisone. When that failed to help, McNair was desperate. Aware that people were trying growth hormone to treat nerve problems and spinal cord injuries, she asked her doctor. "If you can find some, it probably won't hurt you," he said. When she read a small piece on Howard Turney and El Dorado in a Houston paper, she called Turney. At the end of the phone call, she had a job as a nurse at the clinic and a source of the hormone.

By the time we caught up with McNair down in Cancun in 1993, she had been on the hormone for eight months. Two months into the treatment, the hole in her vision had started to shrink. Curious to see if the hormone was responsible, she went off it, and within a short time the hole came back; she resumed treatment, and after five months her eye was completely normal. The cortisone treatment had caused her to gain thirty-five pounds and her face to

become as round as a plate, she told us. Now, her old size-eight figure had returned and her face had clearly regained its contours. In
fact, she looked at least a decade younger than her years. Her
energy level was sky high and she was sleeping at night the way
she did when she was much younger.

It was a happy and exciting time for McNair, but little did she
know it was all swiftly coming to an end. Howard Turney—caught
in an investors' dispute over El Dorado—soon closed the clinic, and
suddenly she was left in a foreign country without a job or anyone
to turn to. She had no money to take growth hormone. In fact she
had little money for anything. It was, she says now, the lowest point
in her life.

"At one point, I was waitressing, eating one meal a day. I was
picking food off the plates for my cat—and something for myself.
This was someone who was a professional woman," she says about
herself. "I have never considered myself a bag lady, but I was one
step away from it. I had a truck, but I had to walk everywhere
because I had no money for gas. So I went through a bad depression and finally an emotional breakdown."

Today it seems almost impossible to square that picture of a
down-and-out near bag lady with the serene, beautiful woman who
still looks more than a decade younger than her fifty-three years.
The El Dorado facility is under new ownership and is now called
the Renaissance Rejuvenation Centre, and she is back at her old job
in Cancun, administering growth hormone. The facility, which has
an outpatient hospital and does a comprehensive physical exam
including blood hormone levels, charges $7,000 a year for supplies
of growth hormone.

She has noticed a number of changes since she returned to the
hormone. For one thing, she has noted, to her delight, that her
metabolism is now back to where it was in her twenties. "When I
hit my forties, I had to change my eating habits to maintain my old
figure. But now on GH, I can eat the way I did in my twenties. My
thighs are also coming back to the thirty-five-year-old level. Everything seems to be tighter. My hands look a lot younger. My nails are
growing." Her digestion has also improved, she says, and she
believes that the absorption from her food and nutrient supplements has improved. For one thing, she says, she no longer has to
take folic acid supplements for leg cramps at night. Another

improvement has been her sex life. "My libido has increased," she says. "Maybe it's just knowing that your body's tighter, and feeling more attractive helps to increase desire. When I was in that depressed state, I couldn't stand facing another day. Now, I'm back to where I was when I was younger. Life is wonderful."

ALLAN AHLSCHIER, M.D. (AGE FIFTY-FIVE)

The first time Ahlschier received a shot of growth hormone, he had a very uncommon reaction. "My hair stood straight up," he said. "What in the world did you give me?" he asked his doctor. It was 1993 and he had gone to El Dorado in Cancun first as a patient and then to stay on as co-medical director. "I was feeling like an old man, like I was going through menopause or something. Maybe it was due to the fact that I was an ex-marathon runner doing about sixty miles a week and probably had injured myself. I thought, I am fifty-three and I'm falling apart."

Today Ahlschier, a board-certified family physician, radiologist, and consultant in anti-aging medicine, is virtually a new person. He has gone from 219 pounds to 267, from a forty-three long jacket to a fifty-three. And it is all solid muscle, he says, no fat. It is not just growth hormone, he says. He went off it for eight months, using growth hormone stimulators. (More about this in Chapter 16.) And he has gone in for bodybuilding in a big way, using 1,000-pound leg presses, for example. But it is the growth hormone, along with testosterone replacement, that allows him to push his workout to the limit. These days he feels like Schwarzenegger. "I shaved my body hair recently so I could see what I looked like and my wife nearly fainted."

It is not just that he looks better, he says; he feels better. "It's an antidepressant. I feel good all the time." And he is doing better. His cholesterol has fallen from 220 to 170, even though he eats eggs for breakfast every day. His blood pressure is normal. "It goes very high when I exercise, which is a stress reaction and appropriate, and then it comes right back down." And his energy level and sex life have reached new heights. His immune system has improved. "I never get sick," he asserts.

Another area in which Ahlschier has noticed a tremendous difference is his vision. "As a radiologist, I look at thousands and thousands of images a day, mammograms, microcalcifications. My whole world is visual." But despite all the strain on his eyes, he has not needed a change in his prescription for either near or far vision. His near vision has improved and his far vision is simply astounding, he says. "I'll be driving along—and this happens every day—and everything will be in perfect focus up to infinity. In perfect, panoramic focus on my cerebral cortex."

Ahlschier, who uses HGH in his practice, says the dosages he uses are a "trade secret," but they don't exceed 2 IU a day, where you start to get noticeable side effects. The cost of the hormone can be cut in half, he says, by buying it directly from pharmacies in Mexico, mostly because of the devaluation of the peso.

"It's the most amazing drug I've ever encountered," he concludes. "My vision is incredible, my growth is incredible. I feel so much younger than I ever did before. You can look at all these drugs. DHEA is a good drug. Deprenyl is useful. But dollar for dollar, growth hormone is the best anti-aging money I have ever spent."

BEDFORD KING, PH.D. (AGE SEVENTY-FOUR)

Soon after his seventy-second birthday, the arthritis that Bedford King suffered from for years had reached the point where he could no longer exercise, work, or drive his car. It was all he could do to get out of bed. "I thought if that was old age, I was ready to go bye." He sold his electronics business of forty-three years and was on the point of giving up. Then he went to the Renaissance Rejuvenation Centre in Cancun and began taking growth hormone. "Within one week," he said, "I knew it was going to help. I was hurting terribly and then my arms just stopped hurting. I just felt so different. I felt like a new man."

Now back in his native Kentucky, he says, "The longer I take it, the better it does. I believe that if I didn't take it, I'd be in bed today." He is back now where he used to be, taking care of himself, driving his car, working out with weights in his bedroom gym every day. A tiny man, only thirty-nine inches tall and weighing sixty-seven pounds, King suffers from scoliosis and a bone disease called osteogenesis. Growth hormone has not reversed his bone loss, but by taking the new drug Fosamax, he has been able to keep it from getting worse. The addition of DHEA, he says, has also helped his memory. "My arthritis is gone except for a pain in my shoulder from the scoliosis. And my muscles are stronger. I am back to lifting weights. I also found that the elasticity in my skin has come back. When I was sick and couldn't move, I thought if I could only get muscle strength back and do what I want to do like I used to. And it worked just like I thought it would."

In fact it has worked so well that King, who holds a doctorate in nutrition, is thinking of taking up an entirely new career in his mid-seventies—commodities trading. "I'm trying to get something that will pay for my growth hormone therapy down the road for as long as I live. I think if you keep yourself well and you got the muscle strength and everything to do it, you can just keep on working and paying for whatever you need in life."

PAUL BERNSTEIN (AGE SIXTY-FOUR) AND SANDY BERNSTEIN (AGE FIFTY-THREE)

Together, Paul and Sandy Bernstein look like a walking advertisement for health, fitness, and beauty. He is Mr. Fitness USA, a title he earned thirty-two years ago. She is a former model and college instructor with a master's in Latin American studies. Paul has a physique that men his age would die for, while Sandy, with her long blond braid cascading down her back, looks about two decades younger than her age.

Yet even this golden couple had started to feel their years. "I had some car accidents, major surgery, was in a cast for four years," says Paul. "And I was coming into my sixties. I didn't feel as well as I used to and the muscle tone I've always had didn't seem to be there although I was still working out. I couldn't retain my energy level. Something was missing." Although he had heard of the healing effects of growth hormone, he was reluctant to try it at first because, like so many people, he thought of it in the same category as anabolic steroids. "I have a very negative reaction toward them," he says, "they drive you crazy and give you steroid rage. You see the effects in the baseball and basketball players who look like football players. My opinion is that they are on steroids."

After researching HGH, Paul and Sandy became convinced that using HGH was not like taking steroids. "This is a hormone that is produced by the body and then the gland doesn't produce it. So to me it is replacing something that has been taken away from me. It is like being a diabetic. It used to be if you were a diabetic, the doctors would tell you were going to die. Then insulin came along. I have two sons that are diabetic. With insulin, they live normal lives. So why shouldn't I turn the clock back if I can replace something that has been taken away from me?"

And that is exactly what Bernstein feels that HGH and DHEA are doing for him. "We all have a little inner feeling when we're younger that disappears. I can feel that coming back now." He feels that his skin tone is better and the severe sun damage sustained from years of lying out on the beach as a youngster growing up in Coney Island "has healed wonderfully."

As a bodybuilder, he is especially impressed by the effect that HGH has had on his exercise regimen. "My muscle tone is almost where it was when I was in my late thirties and my energy level has picked up 20 percent." The last figure is based on the increase in energy and output that he has recorded in his nightly workouts since starting the drug. He has also lost about 4 percent body fat. And his cholesterol level has fallen 30 points since he started HGH. "The reason I am taking this drug is I want to live and be healthier longer and look better longer. People say, 'grow old gracefully.' No one ever grows old gracefully. I don't want to be old. If you can look in the mirror and see a young and vigorous person, you are always going to be young and vigorous."

His wife, Sandy, believes that women may benefit even more than men from growth hormone. "Women have been doing estrogen replacement now for decades. This is another hormone that we can now put back," she says. She is especially excited by the effect that HGH has had on her skin. "It got thicker and firmer." She points to the almost invisible crows' feet at the corners of her eyes. "The little lacework lines are almost gone." She never had a weight problem, she says, but she was still getting bulges on the sides of her thighs. "No matter how much I pedaled, no matter how much I walked, how much I worked out, they would not go away. About six weeks after I started, I noticed that the skin on my face was getting smoother, the lines were going away. Then about four or five months later, Paul pointed out that the bumps on my thighs were going away and now the cellulite is almost gone."

Her cardiac risk factors have also improved. Although her cholesterol levels were not high to begin with, they went down. And her blood pressure dropped from 140 over 90 to 120 over 60—"what it was when I was in college." She too has noticed a big improvement in her energy level and stamina. And along with everything else, their sex life has improved. "Usually, you go to bed to sleep," says Sandy. "It's like, don't bother me. Now it's"—she pauses a moment and grins—"definitely more romantic."

In fact the two of them feel so good, they are starting a new business, opening up health spas. "It had gotten to where Paul didn't want to do anything. He would say, 'I'm tired. I don't feel good today. But now we are ready to meet life's challenges again."

WILLIAM FALOON (AGE FORTY-TWO)

On Thanksgiving weekend of 1991, William Faloon and Saul Kent, vice president and president, respectively, of the Life Extension Foundation, met with four of their members who wanted to try growth hormone after reading about Rudman's study. "They were all older people who were very healthy for their age," says Faloon. "They took tons of vitamins, hormones like DHEA, and they were in good shape but they knew they were going downhill. They made a decision to take a risk."

Faloon put them in touch with Dr. Sam Baxas in Switzerland, who would put them on the hormone, do the blood tests, and ship them new supplies, something no American doctor was willing to do at that point. Within a few months, the program had expanded to forty people, whose ages ranged from thirty-seven to ninety-plus. The foundation does not have a financial interest in growth hormone, he says, but it does have an interest in the results. "Anything else we're involved in, we believe is slowing down and preventing disease and aging," he says, "but the only reversal of aging we're doing is when we make referrals to a number of doctors around the world who sell growth hormone. It is the only anti-aging product I know of where you get immediate gratification."

With other hormone replacement, he says, "you'll get a nice effect in some people and no effect at all in others. Estrogen produces a noticeable benefit in some women and other women just hate it. Melatonin will put you to sleep, which will make you feel a lot better, but as far as actually dealing with weight gain, muscle atrophy, low energy, low libido, neurological dysfunction, skin that's lost its elasticity, only growth hormone will directly reverse those parameters. Within a several months period of time, people feel younger, look younger, think younger, which is why they won't give it up."

There have been some incredible success stories. One fifty-nine-year-old patient, who was on the verge of retiring from his career because of exhaustion and age-related disease before he started on the program, has received a promotion and is now extremely active in his career and private life. A fifty-nine-year-old rancher, who has Charcot-Marie-Tooth syndrome, a rare inherited muscle-wasting disease, was barely able to take a bath without assistance before going on a program of HGH, RNA injections, and DHEA. He can now play tennis and run on his treadmill, and recently married his secretary, who is twenty-five years younger. Lab tests have shown the breakdown of his muscles has not only stopped, but reversed!

Rejuvenation can be a tricky business. Faloon tells the story of an eighty-six-year-old man who had achieved wealth and fame as a result of his scientific discoveries. He called Faloon when he was terrified that his rapidly declining health threatened to rob him of his independence. Within a few months of taking HGH, he was feeling so much better that he had resumed his research full-time. In fact, he was doing so well that his family took away the drug, contending it was no good for him. "He had to buy secondary sources of it and inject himself in the closet where they couldn't see him doing it," says Faloon. "But the real reason the family was taking away the growth hormone was they could see that it was working and instead of his dying in the next five or ten years and leaving them his money, they were afraid they'd have to wait forever."

JOHN BARON, D.O. (AGE EIGHTY-TWO)

At age eighty-two, John Baron looks about sixty and has the endurance of an Energizer bunny. Always a healthy person, which he attributes to the nutritional and vitamin programs he's been on for the last fifty years, he still notices a big difference since he added HGH to his regimen about a year and a half ago. "The most dramatic changes I noticed," he says, "were in my energy, concentration, memory both short and long, and enhanced creativity.

"I can do exercise equivalent to a forty-year-old," says Baron, who directs both the Baron Clinic and the American Institute of Anti-Aging in Cleveland, Ohio. "I have a staff of eight, work an average of eight to ten hours five days a week, and I'm the only doctor here since we're having difficulty finding another one."

Other changes he has noticed include a narrowing of his waist from forty-four inches to thirty-eight, even though his weight has remained stable at 189 pounds, and tighter skin. His youthful face is accentuated by the fact that he has stopped getting gray and his hair has grown back in the spots where he was starting to get bald. His distance vision has improved to the point where he no longer needs glasses for driving. Most surprising, he says, is the return of his sexual prowess. "I'd say it's as good as it was when I was twenty-five."

He is now treating sixteen patients with HGH, ten women and six men between the ages of thirty-five and eighty-four. They have all had dramatic improvement, he says. One interesting observation he has made is that the women become more extroverted, while the men become more romantic.

"The most beautiful thing," he says, "is that you see a new person emerge because of the energy level, the feeling they have about themselves. They lose that sense that they are getting old and they are so happy that they wish they had done this twenty years ago."

VINCENT GIAMPAPA, M.D., F.A.C.S. (AGE FORTY-SIX)

As a pioneer of the new technique of anti-aging plastic surgery, Giampapa actually sculpts the underlying fat and muscle to remove years off people's faces. But his interest in anti-aging is far beyond skin deep. He is director of research at the Longevity Institute International (LII) in Montclair, New Jersey, a unique research and clinical center "which employs revolutionary techniques designed to measure, treat, and retard the aging process and help clients achieve the maximum potential life span for that individual."

As part of his multifaceted attack on aging, he has been treating people for the last two years with hormone replacement, including growth hormone. Most of his patients are healthy individuals between the ages of forty and seventy-five, who are highly focused and motivated to maintain their health, he says.

He will only put people on growth hormone, he says, if their levels of IGF–1 are markedly decreased. "We bring them up to a level of a thirty- to thirty-five-year-old," he says. "At that therapeutic range, we find it not only helps reduce body fat and build muscle mass, but more importantly it keeps your immune response strong. In general, the overall concept of cell rejuvenation, the ability of the DNA to replicate and repair itself, seems to be optimized." He also uses a supplemental, metabolism-enhancing formula that includes amino acids and nucleic acids, which, along with growth hormone, helps supply "the high-test fuel that allows the cell to function at its maximum level for that particular individual at that particular age," he says.

His individualized approach means working with the patient's own pituitary gland as much as possible. "When patients come in, we don't just shotgun them on GH," he says. Instead he starts everyone on his metabolic formula and uses growth hormone releasing agonists, such as L-glutamine, arginine, and ornithine, which costs only about $15 a month. (See Chapter 16.) "If you can stimulate the body to make its own GH, you're in great shape," he says. "If the agonists do not bring the IGF–1 levels into the desired range, he goes to the next step, using stronger stimulants like GHRH (growth hormone–releasing hormone), which is the body's way of stimulating release of growth hormone. Only if that fails to work does he then give growth hormone. "This is a scientific and clinical approach," he says, "where you just don't buy growth hormone and put in your body."

For the same reason, most of his patients who take the hormone use it three times a week rather than six or seven days a week. And after one month of treatment, they go off it for two weeks before starting another round. "We do everything we can to keep their normal feedback mechanism to the pituitary intact."

With that regimen, he says, "the patients do wonderfully and there are no problems with side effects. Their skin looks better, the sex drive is markedly increased, their body composition improves, and there is a subjective increase in their overall sense of well-being." Giampapa, who is helping to develop a sophisticated set of biomarkers of aging, finds that their biomarkers in general improve, and their immune function quantitatively gets better, as determined by laboratory tests.

While growth hormone represents an enormous step in anti-aging medicine, he believes it is just the beginning. His goal, and ours, is to slow the aging process long enough so that "most of us will be around twenty-five years from now to benefit from what medical science will soon be able to do."

Immune Rejuvenation: The Key to Longevity

The stimulation of growth hormone either by injections of HGH or by the use of growth hormone simulators promises to be one of the greatest weapons against disease. It not only rejuvenates immunological vigor, but it is our contention that it actually helps restore the finely tuned balance of the body, correcting the kinds of disturbances in homeostasis that Dilman talked about in Chapter 5.

As we age we become increasingly vulnerable to disease. There are two main reasons for this. First, the body's immune defense system—that is, the ability to fight viruses, bacteria, and cancer cells—declines with age, starting in adolescence. Second, the processes required for the proper maintenance of tissue function, such as the ability to metabolize sugar, handle cholesterol, and clear the kidneys of toxins, become progressively less efficient. The result is the chronic, degenerative, disabling, and ultimately deadly diseases of aging. Growth hormone strengthens the body against disease by acting on both of these factors. And, as we will see in the next chapter, it reverses the decline in physiological processes that protect against disease. First let us take a look at the one of the most exciting experiments ever conducted in immunological research.

GH TURNS BACK THE IMMUNE CLOCK

In 1985 Dr. Keith Kelley, a research immunologist at the University of Illinois at Urbana-Champaign, wondered whether growth hormone played a role in the involution, or shrinking, of the thymus gland. The thymus gland, which is located behind the top of the breastbone, is a primary organ of the immune system. For reasons yet to be determined, at about age twelve, the thymus begins to shrink, until by age forty it is a shriveled shadow of its former self and by age sixty it is difficult to find. The thymus is fundamental to the maturation of the T-cell lymphocytes, which are foot soldiers in the fight against disease. It is the loss of T-cells that makes AIDS patients vulnerable to the devastating diseases that finally kill them. With the shrinking of the thymus comes a rise in the diseases associated with aging, including cancer, autoimmune diseases, and infectious disease (see Figure below). At the same time, there is a decline in the T-cells along with immune factors, such as interleukin 2. In a sense, aging is like a slow form of AIDS.

"The involution of the thymus gland is one of the cardinal biomarkers of aging," says Kelley. He was struck by the fact that just before the thymus starts to shrink around the age of puberty,

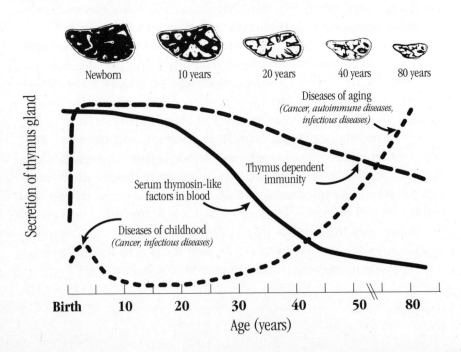

growth hormone is at its highest. This is the time of the growth spurt, when the long bones of the body shoot up. With age, the thymus gets smaller, and growth hormone goes down. Could there be a connection? he wondered.

To test his idea, he injected old rats, whose thymus had almost disappeared, with GH3 cells. These are cells grown in the laboratory that secrete high quantities of growth hormone. To his great delight, the experiment worked. The thymus gland grew back in the old rats so that it was as large and robust as that of young rats. He published his paper, "GH3 Pituitary Adenoma Implants Can Reverse Thymic Aging," in the prestigious *Proceedings of the National Academy of Sciences*.

Before his experiment, he says, "Everyone considered that the thymus went away and you couldn't get it back. But clearly that was incorrect. It was not due to a genetic defect. It was not programmed to go away in the sense that you could not get it back. You could get it back by using a treatment. And that treatment is growth hormone." Kelley also showed that the T-cells from the growth hormone treated rats made more interleukin 2. "The synthesis of interleukin 2 by T-cells in old rats goes down. If you give them growth hormone, it comes back up."

Kelley's work has now been confirmed by Israeli scientists who used bovine growth hormone to reverse thymus shrinking in mice, and similar results have been shown in dogs. Immune activities that growth hormone improves are the manufacture of new antibodies, increased production of T-cells and interleukin 2; greater proliferation and activity of disease-fighting white blood cells; higher activity of natural killer (NK) cells that protect against cancer; stimulation of bacteria-engulfing macrophages; increased maturation of neutrophils, white blood cells that are highly destructive to microorganisms; and increased erythropoiesis, the production of new red blood cells.

GH LINKS MIND AND BODY

Recently scientists have found that growth hormone is made not only by the pituitary gland but by the lymphocyte cells of the immune system. There are receptors in both the immune system and the neuroendocrine system for all the important players in the action of growth hormone, that is, growth hormone itself, growth hormone–releasing hormone, somatostatin, and IGF–1. While the details of how growth hormone from the brain and that produced in the immune system affect each other have yet to be worked out, the important and fascinating thing is that growth hormone serves as a connecting link between the mind and the body. As the researchers conclude, "The activity of all major immune cell types, including T cells, B cells, natural killer (NK) cells, and macrophages, can be altered by GH. We have shown that lymphocyte-derived GH is involved in lymphocytes proliferation and can induce IGF–1 production by cells of the immune system. Taken together, these observations have begun to provide a biochemical basis for one line of communication between the immune and neuroendocrine systems through the actions of GH."

If, indeed, the brain and the immune system talk to each other through the medium of growth hormone, it could explain how the stimulation of GH could have such a therapeutic effect on the mind, the emotions, and the resistance against disease.

GH = ENHANCED IMMUNITY = INCREASED LIFE SPAN

Increased immunity not only means less disease but a longer, healthier life span. In fact, many gerontologists believe that the immune system is the key to longevity. With that idea in mind, let's take another look at the immune system findings in the life span experiment we described in Chapter 5. When Kelley showed in 1986 that growth hormone could reverse the shrinking of the thymus, it greatly excited a young graduate student, David Khansari, who was getting his Ph.D. at the University of Illinois, where the work was carried out. If growth hormone was able to rejuvenate the immune system, he wondered, might it also extend the life span? Several years later when he had his own laboratory at the North Dakota State University in Fargo, he and Thomas Gustad carried out the experiment that showed a one-third extension of average life span in older mice treated with growth hormone.

As mentioned earlier, the immune system declines dramatically in older age, opening the door for infectious diseases and cancer. But the GH-treated old mice actually improved the production of some immune factors, including interleukin 1, immunoglobulin G, and tumor necrosis factor. These results, say the investigators, support the idea that the T-cells in aged animals (and presumably humans) are not inherently defective, but can be reactivated by growth hormone.

Why did the treated animals live longer? The studies on immune function indicated that a number of immune factors called cytokines increased with immune therapy. As the researchers note, "Cytokines play a central role in all aspects of immune response. In addition they are a means of communication (regulatory) between the central nervous system and the immune system." Like Kelley, they found that growth hormone increased the level of the cytokines, interleukin 1 and 2, and tumor necrosis factor. In fact, the researchers found no difference in the production of these factors between the old treated mice and young mice.

Although not every immune function that the researchers tested improved with growth hormone, they still believe that the immune system played the major role in the longer survival of the growth hormone–treated animals. "One theory is that the growth hormone rejuvenates the immune system," says Gustad. "It halts or at least slows the breakdown of the thymus itself and therefore cell-mediated immunity may be prolonged and along with that, such things as improved immune surveillance. We felt and still do that [the extension of the life span] may be due to a rejuvenation of the immune system or at least a halting of the natural decline of the immune system."

A BALANCING ACT

Kelley agrees that growth hormone may rejuvenate the immune system and not just by bringing back cells that have disappeared with age. "Our old idea that the cytokines decline with age and growth hormone brings them back turned out to be simplistic." While some immune factors like interleukin 2 decrease with age, he points out, others like interleukin 4 go up. Interleukin L4 can act as a suppressor of the immune system. "The immune system turns out to be very complicated," he says. "But the take home message is that with age, there is a dysregulation. Everything gets out of balance rather than everything declines. What immunologists are trying to do is get things back into balance. And I think in a broad sense that is what growth hormone is doing. It is letting those cells talk to each other the way that they used to."

REJUVENATING THE BONE MARROW

Imagine if you could replace your old blood with new young cells. It would be like prolonging the life of your car by changing the oil every few months. Kelley's most recent research shows that growth hormone replacement therapy may be doing for your blood and immune system what a change of oil does for your car's engine. Only in this case, it is the bone marrow that is being replenished. In effect, you are giving yourself a bone marrow transfusion.

The bone marrow of the long bones is the birthplace of all the cells that ultimately become blood cells in the body, including the red blood cells and the white blood cells of the immune system. These are the stem cells that give rise to the blood-forming elements called progenitor cells that eventually develop into the red blood cells and the white blood cells of the immune system. The stem cells also give rise to the T-cells that mature in the thymus. Kelley found that the bone marrow of both old humans and rodents were becoming depleted of their progenitor cells, the mother cells that actually give rise to all the blood cells. He calls this phenomenon hypocellularity, a lack of cells. But this disappearance, like the thymus shrinking, turned out be reversible. When he gave growth hormone to the old rats, the progenitor cells came back.

Kelley thinks that the problem lies not in the stem cells, which can outlast the human life span, but in the environment of the bone marrow where these stem cells sit. "Somehow the hormones act to overcome a block. The thymus involutes, the cellularity goes away in old folks. You give them growth hormone and they both come back. We don't know exactly what happens. Magic occurs. But where we once thought the action was all in the thymus, now we think that the action is in the bone marrow.

"I think that another target beside muscle cells and fat cells for the aged is the immune system," he goes on. "For me, it is just so fascinating to come to the realization that hormones, which control everything else we do, do something for the immune system. I think that the fact that growth hormone affects the immune system is a major advance in terms of how the body works in aging."

HUMAN IMMUNE BIOMARKERS

Vincent Giampapa has been looking at immune system function in the patients he has treated with growth hormone. These are immune biomarkers, that is, biological measurements of aging. With HGH, the T-cell lymphocyte function improves along with increases in red blood cell formation, synthesis of antibodies, cell division of lymphocytes, and two cancer-fighting factors—natural killer cells and tumor necrosis factor alpha. Other studies reveal a growth in the size of the human thymus and DNA synthesis of the cells in the thymus.

"What growth hormone does is jump-start the older cell," explains Giampapa. The cell goes through different stages of cycles of division, the G-0 stage, which is quiescent, and the G-1 stage, where it divides. But as you age, the cells stays longer in the quiescent stage and can accumulate genetic damage that is not repaired until it divides. "Growth hormone pushes the cell so it goes into the repair and reproductive stage more quickly," he says.

The fact that growth hormone rejuvenates the immune system has incredibly far-reaching implications, according to Greg Fahy, Ph.D., a prominent researcher in cellular physiology at the Naval Medical Research Institute in Bethesda, Maryland, and a respected authority on anti-aging measures. "With age, the incidence of cancer, infectious disease, and other maladies goes up," he said at a meeting of the American Academy of Anti-Aging Medicine. "It has been shown in animals and with very strong suggestive evidence in humans that this can be prevented or actually reversed. For the first time we have a chance of maintaining an immune system at the age of 80 which is similar to what we had at 20." But restoring the immune system is not just bringing back the ability to fight disease, he pointed out. "If you restore immune function, your ability to make DNA, to have normal cell division, to have normal insulin sensitivity, to have normal thyroid hormone levels and other things such as normal populations of certain molecules in the brain that change with age, all these things are restored by an improvement in the immune system."

Even more exciting, he said, "It is clear that many parts of the body machine are interlinked and if you can straighten out one part, you may find that the defect in that part was actually the reason for a defect someplace else. Which means that aging is much more simple than we had any right to expect or hope. If we can find the few key cogs in the machine of the aging process and set them right, the rest of the machine may follow along. This is a very exciting concept and seems to be borne out by what we know today."

Fahy strongly believes that growth hormone is one of the most promising treatments for human health and reversal of degenerative disease. "I think that a lot of aging can be explained by deficits in growth hormone. I don't think that even people working in the field see the whole picture. They see only the superficial things they are aging."

He speculates that drugs that are used for anti-aging purposes, like deprenyl, co-enzyme Q10, and centrophenoxine, or melatonin may simply be patching up the holes left by the withdrawal of growth hormone. For instance, he points out, deprenyl, an anti-Parkinsonian drug that extends life span in rodents, slows the loss of the brain chemical dopamine, and there is some suggestion that dopamine deficiency may play a role in the slowdown of growth hormone with age. Melatonin, which is produced by the pineal gland and has been touted for its anti-aging effect, starts falling in puberty. "The drop in melatonin," says Fahy, "may trigger the fall in growth hormone. And it may be that the reason melatonin is good for you is that it stimulates growth hormone release." Co-enzyme Q10, a naturally occurring, vitaminlike constituent of most cells, which has also been shown to extend life span in animals, may slow the fall in growth hormone. And centrophenoxine, a European anti-aging drug and brain stimulant, restores protein synthesis, which goes down when growth hormone declines. "If you put all the pieces of evidence together, it suggests that growth hormone is even more central to the aging process than most people think," says Fahy.

There is no doubt that immunity, aging, and longevity are closely linked. And the common thread running through all of them may be the rise and fall—and rise again—of growth hormone.

8

The Medical Miracle

DISEASE AND HOMEOSTASIS

Replacing or stimulating the release of growth hormone can maintain your body in the healthful state that is enjoyed by the vast majority of young people. It accomplishes this medical miracle by helping the body to maintain its internal balance, which is known as homeostasis. The *American Heritage Dictionary* defines homeostasis as "The ability or tendency of an organism or a cell to maintain internal equilibrium by adjusting its physiological processes." In other words, the living organism is a self-adjusting machine. The ability of the body to shift its equilibrium is due mainly to the action of hormones. The hormones work on a feedback system like the thermostat on a furnace. A temperature change in a room that differs from the thermostat setting signals the air conditioner or the furnace to turn on or off until the desired setting is reached. In the same way, if we are too hot, the sweat glands allow us to perspire and cool off, to maintain a stable internal body temperature. As we age, the hormones become less efficient, the body has more and more difficulty maintaining homeostasis, and disease results. Or to put it another way, much of what we call disease is a disturbance in homeostasis. By improving the metabolism and correcting homeostasis, growth hormone replacement therapy can both prevent and treat diseases of aging.

SWEAT EQUITY

Talking about sweat glands, one of the problems in people with growth hormone deficiency is that they sweat less. In a study at the Kommunehospital at Aarhus, Denmark, Dr. Jens Sandahl Christiansen and his colleagues found that growth hormone replacement brought sweat production back to normal. While this might seem a trivial problem, as the investigators note, "Our findings indicate that GH might have a role in thermal regulation, via an effect on sweating and that some of the positive effects reported following GH replacement might be due to a normalization of sweat secretion," such as the enhanced ability to exercise.

HELP FOR AIDS

The first new medical indication for HGH in the United States other than for growth problems in children will most likely be for AIDS. Clinical trials are now going on in a number of medical centers across the country using Serostim, a form of recombinant human growth hormone made in mammalian cells by the Swiss-based company Serono. It is used to reverse a pernicious form of weight loss seen in later-stage AIDS, called "wasting." Unlike the loss of fat that comes with undereating, wasting, which also occurs in later stages of cancer, is unintentional loss of weight, in which the lean body mass, the vital muscle, bones, and organs are withering away. (The disease is known as "thin" in Africa.) Many AIDS experts believe that losing lean body mass contributes to the immune dysfunction and makes it harder for people to fight off life-threatening infections. And the wasting process itself, if it goes beyond about 33 percent of ideal body weight, is incompatible with life, as known from the experience of concentration camp victims.

In a study of 178 patients, at a number of medical centers, with AIDS-associated wasting, defined as having lost at least 10 percent of body weight, those people who received growth hormone gained an average of 6.6 pounds of lean body mass while losing an average of three pounds of fat after three months of treatment. The therapy also resulted in an improvement of their endurance and quality of life.

"At the moment there isn't another therapy in the late stages of the disease that has had this kind of effect on the lean body mass," says Morris Schambelan, M.D., professor of medicine at the University of California in San Francisco, one of the principal investigators. "There are therapies that can increase weight, such as an appetite stimulant called megestroacetate, but that weight gain tends to be primarily fat. The advantage of growth hormone is that it causes lean body mass to increase."

Schambelan says that some of the patients are still coming in since the original study ended three years ago. "Obviously these are the people who are still alive and doing well and the people who didn't do well are no longer there, so you can't talk about the average effect. I think that the anecdotal experience is that for the people who continue to take the drug, who continue to eat, and don't get serious opportunistic infections, they have had a very robust response. The medication has increased the benefits. The gain in lean body mass that they had in their three months of the study has doubled or tripled over the next year or two."

Growth hormone does not cure AIDS but there is some indication that it might lengthen people's lives. The increase in lean body mass may make them less vulnerable to infection. When given with antiviral drugs like AZT, growth hormone did not cause any increase in the amount of HIV and it stimulated the formation of new red blood cells, which are depleted by AZT. Even more exciting, it appears to reduce the incidence of AIDS-associated infections, such as polycystic pneumonia and Kaposi's sarcoma. As we discussed in the previous chapters, growth hormone can rejuvenate the immune system. If it can actually strengthen the ability of HIV-positive patients to resist infectious diseases, it may improve both the quality and quantity of their lives.

WASTING IN THE ELDERLY

Growth hormone also reverses a kind of wasting due to malnutrition in the elderly. For reasons that are not well understood, the appetite of old people often fades away to nothing, which has been called the "anorexia" of aging, says Fran Kaiser, M.D., associate director of geriatrics at the Geriatric Research, Education, and Clinical Center of the Veterans Administration in St. Louis. Others may eat well, but fail to thrive. As many as 65 percent of elderly who are hospitalized and between 30 and 60 percent of those in nursing homes may be malnourished. "Recent studies suggest that age-related reductions in the natural production of growth hormone contribute to malnutrition and frailty in the elderly," writes Kaiser. "If this proves to be true, then recombinant human growth hormone could become a significant therapy for these patients."

In a double-blind clinical trial of ten malnourished elderly men with no diagnosable cause of weight loss, Kaiser found that at the end of twenty-one days, the five who received growth hormone ate more, gained weight, increased their muscle mass, and were excreting half the nitrogen of the control group—an indication of muscle-wasting—which showed that their protein metabolism had improved. On the other hand, those who received placebo injections continued to lose weight and take in even fewer calories, even though they were urged to eat by hospital personnel. So not only did growth hormone improve the metabolism and physical well-being in this elderly group, but it obviously made them feel better because they began to eat again.

REVERSING HEART DISEASE

The number one cause of death among both men and women in the United States is heart disease, accounting for more than one-third of all deaths. Cardiovascular disease and heart attacks account for one and a half million deaths a year. Stroke is in third place, with chronic obstructive pulmonary disease in fifth. There is increasing evidence from a number of studies around the world that growth hormone therapy can both prevent and reverse these age-associated killer diseases.

The central role of growth hormone in cardiac disease can be seen by what happens when people have too little or too much of it. Both those patients who are growth hormone–deficient and those who have acromegaly—a disease due to overproduction of growth hormone—die of increased heart disease. According to Bengtsson of the University of Göteborg, "there is a U-shaped relationship between growth hormone and cardiovascular disease. Too much and too little is dangerous." But when the body has the proper amount of growth hormone that it needs, the effects on heart function are dramatic.

Deaths from Cardiovascular Disease (in 333 HGH-Deficient Patients)

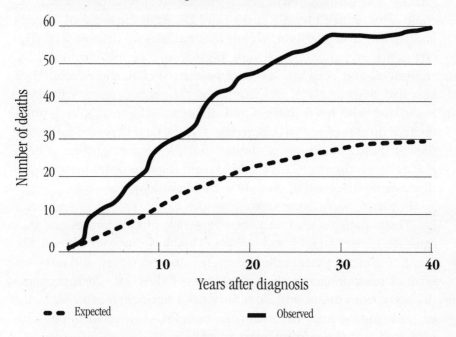

Years after diagnosis

● ● Expected   ■■■ Observed

Adapted from: Rosen, T., and B.A. Bengtsson, *Lancet* 336 (1990): 285–88.

In his study of 333 patients with severe growth hormone deficiency at Sahlgrenska Hospital, Bengtsson found that they had twice the mortality from heart disease when compared with people matched for age and sex (see figure above). To zero in on what could possibly account for this unexpected rise in cardiac deaths, he and his colleagues studied the cardiovascular risk factors in one hundred of the surviving patients.

The first thing they looked at was body composition. The men were more than sixteen pounds overweight and the women more than seven pounds heavier than normal. But even more significant, they had a higher fat-to-lean ratio than average. And most of their fat was concentrated in the abdomen, the area known to raise the risk of heart attack.

Second, he looked at their cholesterol level. He found that the patients had normal cholesterol levels, but when he split the cholesterol between LDL cholesterol and HDL cholesterol, the levels told a very different story. LDL, or low-density lipoprotein, is the bad kind of cholesterol, which clogs the artery walls and can cause heart attacks and strokes. HDL, or high-density protein, is the good cholesterol, which literally sucks out LDL from the walls of arteries and carries it to the liver, where it is harmlessly disposed of. The GH-deficient patients had very high levels of the harmful LDL cholesterol and very low levels of protective HDL cholesterol. They also had high levels of the fats known as triglycerides, which are associated with heart disease and diabetes. "This was a very unfavorable lipid profile," says Bengtsson. In fact, he says, the lipid profiles of these patients were similar to those of normal men at high risk of heart disease. "This could explain the double increase in cardiovascular disease that we saw in these patients," he says.

But there were other factors besides cholesterol level and body fat. These patients also had increased risk of clot formations that can cause heart attacks and strokes. This was due to increased fibrinogen, a blood protein that promotes clot formation, and a higher level of plasminogen activator inhibitor–1 (PAI–1), which prevents the body from dissolving clots. Increased fibrinogen is a risk factor in both stroke and heart attack, and PAI–1 is associated with increased risk of recurrent heart attacks.

The growth hormone–deficient patients were also more apt to have higher blood pressure than normal, and increased insulin resistance. The latter finding came as a surprise since children who are short due to growth hormone deficiency usually have increased sensitivity to insulin and tend toward hypoglycemia. And growth hormone is known to block the effects of the insulin on the body. But when Bengtsson and his co-workers measured the uptake of glucose on the GH-deficient patients, they found just the opposite. The patients had decreased insulin sensitivity and were actually insulin resistant.

Summing up, Bengtsson says, "These patients had increased cardiovascular mortality. They have increased risk factors in the form of increased body fat, low HDL, high LDL, high triglyceride concentration, high fibrinogen, high plasminogen activator inhibitor–1. And they have increased peripheral vascular hypertension and are insulin resistant. These patients have a lot of cardiovascular risk factors that could explain their increased cardiovascular mortality."

What happened when these patients were given growth hormone replacement? First, it had a profound effect on their body composition, trimming their fat and bulking their lean body mass. Bengtsson also made a remarkable discovery that showed that growth hormone therapy actually clears the deep fat out of the abdomen. Second, it moved the blood cholesterol in the desired direction—raising the HDL and lowering the LDL. After six months of treatment, growth hormone reduced the diastolic blood pressure (the lower number on a blood pressure reading) by about 10 percent, although it did not change the systolic pressure. "This was in total contrast to what was thought before," says Bengtsson. "Everyone thought that growth hormone would increase blood pressure and we showed the opposite."

HGH REPLACEMENT AND DIABETES

There has been some concern that giving HGH could make diabetes worse, since growth hormone is known to antagonize the effects of insulin. But recent studies have shown that far from worsening diabetes, it may actually help. Previous studies, including Rudman's, showed that growth hormone caused increased insulin resistance, making blood glucose rise, and contraindicating its use in diabetic patients. Bengtsson showed the same thing in a short-term study of six weeks of growth hormone treatment. But after six months, he found, the insulin sensitivity returned to where it was before the patient started treatment. In other words, growth hormone did not increase blood glucose or make the patient diabetic. It may be, says Bengtsson, that the improvement in body composition after six months offsets any negative effect that growth hormone has on insulin.

There are some indications that growth hormone may even improve diabetes. Blackman, who is looking at the effects of growth hormone on blood glucose as part of his NIA study at Johns Hopkins, says, "We postulate the case that when elderly people are in a low state of growth hormone, they have more intrabdominal fat, their insulin acts less well, and they have a tendency to higher blood sugar. If they are given growth hormone replacement carefully rather than excessively, their blood sugars should improve rather than worsen."

And in a conversation with Greg Fahy, Ph.D., of the Naval Medical Research Institute, he revealed that his own unpublished experiments show that growth hormone is extremely beneficial for diabetes and he may have discovered a method to potentially cure adult-onset diabetes via growth hormone. He has filed patents for the treatment of diabetes with HGH.

HGH AND ATHEROSCLEROSIS

Bengtsson expects that future studies will show that growth hormone will actually reverse atherosclerosis in the coronary arteries. The latest view of atherosclerosis is as a metabolic disease, he says. While most people focus on cholesterol plaques in the artery, the real action takes place in the liver. This is the primary organ for disposing of cholesterol by transforming it into bile acids or shipping it into the gallbladder and from there to the intestines, where it can be excreted from the body. Growth hormone enhances this metabolism by increasing the number of LDL receptors in the cells of the liver that remove LDL cholesterol from the circulation. "This is the way to clean up the circulation from cholesterol," says Bengtsson. In other words, growth hormone acts directly on the root cause of most heart disease and heart attacks.

GROWTH HORMONE IMPROVES HEART-LUNG FUNCTION

HGH makes the heart a better pump and the lungs a more efficient air-moving machine. Studies in a number of countries indicate that growth hormone multiplies the number of heart cells and improves cardiac function in adults with pituitary deficiency. In Italy, growth hormone treatments increased the thickness of the left ventricle wall in patients and made their hearts work better. It also benefits heart-lung function by enhancing the ability of patients to exercise, raising their maximum oxygen uptake, and increasing the stroke volume and cardiac output. And it improved one of the most significant measurements of lung function—forced expiratory volume (FEV1), or vital capacity. FEV1 is the ability to force the air from lungs in one second. Vital capacity is the total volume of air expelled from the lungs by breathing out as hard as one can after breathing in as hard as one can. Vital capacity declines in a straight line with age and is considered one of the best predictors of how long a person may be expected to live. While one study found no improvement in vital capacity after six months of treatment with growth hormone, Bengtsson and Rosén found that *one year* of treatment reversed the deterioration in lung function by raising the vital capacity in pituitary deficient patients. This indicates that growth hormone treatment "may have long-term beneficial effects on the pulmonary function." And, we might add, it also shows that HGH alters the downward course of one of the most significant biomarkers of aging and longevity.

REVERSING LUNG DISEASE

In a short-term study of patients with chronic obstructive pulmonary disease (COPD), David Clemmons, chief of endocrinology at the University of North Carolina School of Medicine in Chapel Hill, found that three weeks of growth hormone treatment turned around the decline in breathing capacity. All the patients were in the later stages of their disease, were underweight, and had considerable loss of the ability to expand their lungs during breathing.

How well you can breathe depends on how much negative pressure you can generate. The chest functions like a bellows, so that when you take a breath, it generates negative pressure inside the chest cavity and the air flows passively from a region of greater pressure to lesser pressure. A normal person breathing in at maximum force generates 100 millimeters of mercury negative pressure. The patients with lung disease in the North Carolina study were at 25 millimeters of negative pressure, or one-fourth of normal function. They had similar declines in the ability to force air out of their lungs. If the air is not expelled from the lungs, it remains as dead air space in the lung sacs. This lessens the amount of negative pressure that can be generated, more air stays as dead air space in the lungs, lung capacity declines, and it gets get harder and harder to force the air into and out of the lungs. Eventually the patient dies of metabolic complications associated with insufficient respiration—in other words, suffocates.

In current medical practice, there is no effective way to stem the decline of lung function in patients with emphysema. There are only ways to ameliorate the disease, like cutting out the nonworking lung tissues or having the patient do exercises to increase the capacity of the chest—until now. Very exciting, early-stage research shows that growth hormone regrows lung tissue and restores respiratory function. In the North Carolina study injections of growth hormone raised the maximum inspiratory force by 10 to 12 millimeters on average. The hormone treatment also increased their maximum expiratory force. "If you are as impaired as these patients, going from 25 millimeters to 38 millimeters would be quite a help," says Clemmons, who believes that growth hormone is a promising treatment for emphysema and other chronic obstructive pulmonary diseases. "Anything that would increase inspiratory and expiratory force potentially can be helpful."

REVERSING CARDIAC FAILURE

Growth hormone may protect against chronic heart failure, according to several compelling studies. In a 1996 study published in the prestigious *New England Journal of Medicine*, the researchers gave growth hormone to seven patients (five men and two women) who had moderate to severe heart failure. The treatment consisted of 4 IU of GH given every other day for three months along with the standard medical therapy for heart failure. Growth hormone increased the thickness of the left ventricular wall, enhanced the ability of the heart to contract and pump out blood, reduced the oxygen requirement of the heart, and improved exercise capacity, clinical systems, and the patients' quality of life, according to the study. Since there are no good treatments for reversing heart failure, an editorial in the same issue of the journal called the short-term study "exciting and dramatic," and said that it was "cause for cautious optimism."

In another recent study, researchers at Genentech experimentally induced heart failure in rats by cutting off the circulation to the heart from the left coronary artery. Four weeks later, the researchers treated the animals with growth hormone for fifteen days. The GH-treated animals showed enormous improvement with increased cardiac output, stroke volume (the amount of blood the heart pumps with each beat), and ability of the heart to contract. Based on their results, the researchers are planning to do a clinical study of growth hormone treatment in patients with congestive heart failure.

Several case studies show that growth hormone therapy can have a remarkable effect on GH-deficient patients with heart failure. In one case in Italy, a woman was diagnosed with cardiac failure twenty years after developing pituitary disease that had gone unrecognized. She was given cortisone and replacement of thyroid hormone T4, along with therapy for her heart condition, including digoxin. When she did not improve, her doctors added growth hormone injections. Three months later, the therapy had regrown the wall of her heart muscle and strengthened its pumping ability. The patient improved dramatically.

In an even more dramatic case study, a patient with Cushing's syndrome, a hormonal disease, at St. Thomas' Hospital in London had congestive heart failure that failed to respond to conventional medical treatment. His situation had deteriorated to the point that he had become a candidate for heart transplant. Since no donor hearts were available, Dr. Peter Sönksen, a pioneer in growth hormone research, decided to try growth hormone replacement. The treatment increased the ability of the heart to contract, and his peripheral circulation improved so much that the patient no longer required a transplant. "The cardiac effects are marvelous, thrilling, extremely exciting," says Bengtsson. In GH-deficient patients, HGH increases the thickness of the left ventricular mass, raises the cardiac output, strengthens the stroke volume, and improves the pumping of the blood out of the heart. Yet all these positive benefits do not have the drawback of increasing blood pressure. Indeed, as mentioned, the diastolic blood pressure goes down. This means that somehow the body compensates by decreasing the peripheral resistance of the arteries and capillaries throughout the body to the blood flowing through it, allowing the blood pressure to remain low, he says. Although his studies are yet to be published, he indicated that they are extremely promising, pointing to growth hormone as a future treatment for cardiac failure.

REVERSING HEART ATTACKS AND STROKES

But the most mind-boggling effect of growth hormone may be its ability to stop cell death after a heart attack or stroke. Earlier, in Chapter 5, we mentioned that growth hormone had the power to stop apoptosis, programmed cell death. Now studies on animals show that it can protect the cells of the heart and brain from dying after a heart attack and stroke.

In one study the researchers produced heart attacks in rats and then immediately gave them one of three treatments—growth hormone, IGF–1, or a placebo (saline solution). When they autopsied the animals, they looked at the cardiac tissue affected by the heart attack. Both the groups with growth hormone and IGF–1 had far less tissue death than the group on placebo. In a similar study in animals, a group in Auckland, New Zealand, found that IGF–1 treatment rescued brain cells when given *after* the injury to the brain had occurred. (For details of this study, see Chapter 12.)

The implications of these studies are staggering. The reason that a heart attack or stroke is so devastating is that cells die. Tissue death is forever. If an area of heart tissue dies, the heart function is irreversibly compromised. It is like trying to drive an eight-cylinder car on two cylinders. If tissue in the brain dies because of a stroke, the results can be even more horrendous depending on the area affected. The patient may never regain speech, movement in arms or legs, or the ability to think clearly. By preventing cell death after a stroke or heart attack, growth hormone could prevent the irreversible effects on the heart and brain.

But the beneficial effects of growth hormone may not even stop there. Cell death is a feature of degenerative brain diseases of aging like Parkinson's and Alzheimer's. There too, growth hormone along with other hormone replacement therapy may stop cell death and stimulate regeneration of dying cells. In Chapter 12, we will present some truly amazing results of growth hormone in treating these diseases.

GROWTH HORMONE AND OSTEOPOROSIS

Some 100,000 people a year die from complications of hip fractures. Bones become more fragile with age, and break more easily. The bones of all of us, not just women after menopause, lose their mineral content and density, becoming more porous. And the bone loss occurs throughout the body in the bones that anchor the teeth, in the hips and legs, in the spine. The loss of bone and the loss of water in the intervertebral disks with age is why we shrink. Take a look at your height on your driver's license. Now measure yourself.

Chances are that if it has been a few decades since you got your license, you are already shorter. It now appears that the withdrawal of growth hormone from the bones as you age may be a major cause of bone loss and the resulting fractured hips.

To understand the role of growth hormone in bone, let's look for a moment at how bone is formed. The bones of our body are constantly being renewed. As new bone is formed, old bone is being resorbed. Without this remodeling process, we would turn into bony, overgrown giants. And indeed this is what is seen in acromegaly, where people turn into hulking seven- and eight-footers with huge hands and feet because of the oversecretion of growth hormone. The cells that form new bone are called osteoblasts, while the cells that break down bone in the remodeling process are called osteoclasts. There is a steady increase in bone mass from birth to about age thirty as the rate of osteoblasts exceed that of osteoclasts. After age thirty, the reverse is true, with the activity of the osteoclasts that destroy the bone exceeding the activity of the osteoblasts that make new bone. In men over age fifty, new bone formation averages about 2 percent a year, while bone resorption is 2.2 percent. The result is a steadily creeping loss of bone, or negative bone balance. In elderly people, there is a strong correlation between the loss of teeth, the atrophy of the jaws, and the degree of skeletal osteoporosis.

Growth hormone seems a natural for bone growth since its original therapeutic use was to grow the skeletal bones of young children who were deficient in the hormone. But the results in adults have been inconsistent. Rudman found that six months of treatment increased the density of the lumbar vertebrae by 1.6 percent in the men over age sixty-five. But he found no increase in the bone density of the radius bone in the arm or the femur of the leg. In his paper, he noted that longer periods of administration were needed to tell whether GH could reverse bone loss. Several groups have also reported an increase in bone mineral content, including osteocalcin and alkaline phosphatase, which indicate an increase in bone formation. But short-term treatment has not improved osteoporosis and in some cases, it even resulted in loss of bone density.

But an unpublished study done at Bengtsson's laboratory led by Gudmundur Johannsson showed that growth hormone replacement results in a *positive bone balance*. The study involved twenty-four men and twenty women between the ages of twenty-three and sixty-six who were severely deficient in growth hormone. These are people who have the bone content and density of old people. They are also twice as likely to fracture their bones as the normal population.

After two years on growth hormone, they had significant increases in the density of the bones that form the hip joint and the vertebrae of the lower spine. They also showed an increase in calcium, osteocalcin, and two types of collagen, which are markers of bone formation. The people who benefited the most were those whose bones were in the worst shape to begin with and who were most vulnerable to fracture. And the bones are continuing to get stronger after the two-year study, with a further increase in bone density expected, say the researchers. Since bone density in the regions of the hip and lumbar spine are directly related to risk of fracture in these areas, the researchers estimate that they have reduced the possibility of fracture so that it now equals that of normal healthy controls!

The reason that other researchers haven't gotten good results, says Bengtsson, is simply that they haven't carried out the studies beyond one year. Growth hormone, he says, restarts the bone modeling process, and in the first year, more bone is resorbed than is made, with the result that there is a slight decrease in bone mineral content and bone mass. But, after the first year to year and a half, the reverse is true; the bone mineral content increases along with bone mass.

Growth hormone strengthens bones in a number of ways, say the Swedish researchers. Research on cultured bone cells in the test tube shows that growth hormone acts directly on the osteoblasts to stimulate new growth. It also enhances the availability of vitamin D-3, which, combined with calcium, has been shown to increase bone mineral density and reduce fractures among elderly women. The other known benefits of growth hormone replacement, such as a increased sense of well-being and a greater ability to exercise, may also play a role by encouraging people to do more weight-bearing activity, such as walking.

Will growth hormone reverse the shrinking of aging? "Possibly. We don't know," says Bengtsson. "All our patients increase a little in height," he says. The champion grower from his studies is a woman with growth hormone deficiency, who grew four centimeters, or more than one and a half inches! The growth occurs in the intervertebral disks of the spine, which are the shock absorbers between the bones of the spine, and which are mostly water. Growth hormone does two things: It increases bone mineral content in the disks and it increases protein synthesis. The new, healthy protein holds on to the water which oozes out with age. The result is that with GH replacement, the spines get longer. And the increased bone density means stronger spines.

There are even hints of the bone-building effect of growth hormone in a one-week study of postmenopausal women, where the level of osteocalcin needed for new bone increased, although the osteoclast, or breakdown of bone, continued at the same rate. "The difference in magnitude between osteoblast (bone-building) and osteoclast stimulation suggests that GH should induce a positive bone balance with every turnover cycle," write William H. Daughaday and Steven Harvey in the book *Growth Hormone.* They believe that "bone turnover may also prevent excessive aging of osteocytes (bone cells) and thus contribute to the maintenance of healthy bone tissue."

A definitive answer on whether growth hormone is effective in treating osteoporosis in the elderly awaits the results of placebo-controlled clinical trials now being carried out.

"We believe strongly that GH is important for adult bone mass and that this mechanism in the future can be used for the treatment of osteoporosis," Bengtsson says. At the present time, he points out, the present treatments for bone loss, estrogen and powerful new hormonelike drugs such as Didronel and Fosamax, turn down the bone metabolism and freeze bone loss. But, he says, they do not form new bone. "If you use an anabolic agent like growth hormone," he says, "you activate the growth hormone receptor in the bones." This, he says, is closer to what nature does in the body.

In our opinion, the best time to stem the loss of bone is before osteoporosis sets in. This can be done by following a regimen to stimulate growth hormone, which we describe in the second half of this book.

THE TISSUE HEALING POWERS OF GH

Think about your children or yourself when you were a child. Remember how many times you fell down and scraped your knees?

Your arms and legs always seemed to be wearing red badges of courage. But, practically overnight, a scab would form and the wound would soon disappear, leaving no trace. When we're young, cells regenerate and repair faster, wounds heal more quickly, bones knit more speedily. All these processes are under the control of growth hormone and growth factors. The use of growth hormone and growth factors (see Chapter 10) won't restore the repair rates of an eight-year-old, but it can dramatically accelerate the healing process in those age fifty and above.

One of the first things that people on growth hormone replacement notice is its almost miraculous healing powers.

As one doctor who uses it on himself as well as his patients says, "it just seems to go to the part of the body where it is most needed." One sixty-three-year-old family physician in Arizona reports that he recently started using growth hormone on himself after he fell off his bike and fractured his elbow. Six weeks later he was doing pushups. His astonished orthopedist told him that at his age, it generally takes five months to heal. The Arizona physician was so impressed, he applied to the FDA to do experimental studies on HGH in the elderly population in his community.

But for sheer drama, almost nothing beats what happened to Dr. Lord Lee-Benner, who believes that the hormone literally saved his neck. Two years ago, at the age of sixty, Lee-Benner decided to become active again in a sport that had once been his life's passion: polo. It had been ten years since he played, but he was soon doing well enough to play in a seniors' tournament. His team swept the tournament and he was voted most valuable player.

Ironically, it was not tournament play but a small game with friends that ended in the worst accident of his life. His horse suddenly reared straight up and fell over backward, pinning Lee-Benner under him. He was knocked out, and when he came to, he saw the horse rearing up and falling down on top of him a second time. Again, he lost consciousness. He awoke again to find that he had suffered a concussion, his right leg was broken above the ankle, and he could hardly breathe. When he tried to move, the searing pain in his back stopped him.

As a polo player, he had sustained almost every kind of injury, he says. "I've had the breath knocked out of me, had two broken necks, eighteen rib fractures, separated shoulders, broken arms, broken all the fingers on both hands, and have three pins on both elbows. I've had three horses fall on me and had horses die under me. But I never felt anything like this."

It turned out that Lee-Benner had a collapsed lung, a broken leg, and a contusion of his spinal cord between the T–4 and T–7 vertebrae. "I am a neurologist," he says, " and I know that a swollen spinal cord has a grim prognosis. I should have been paralyzed."

Aware that there was new research showing that growth hormone stimulates growth factors in the spinal cord and can regenerate peripheral nerve tissue, he began taking growth hormone almost immediately after the accident. In six weeks, his leg fracture had healed completely and his lung was restored to full function. The swelling on his spinal cord also disappeared. "My MRI was perfect," he says. "You would never know that there had been an injury."

The healing properties of growth hormone have already been used for patients who are injured, severely burned, recovering from surgery, or severely malnourished. In animal and human studies of wound healing, injections of growth hormone stimulated the formation of collagen (the glue that holds tissue together), increased the tensile strength of wounds, and made the wound close faster. Growth hormone and growth factors have also been used successfully to treat leg ulcers. Studies in rats showed that it healed incisions faster, strengthened bones after fracture, and helped reverse muscle atrophy and weakness after surgery. In one study of burned patients, growth hormone markedly stimulated protein synthesis needed for healing in the whole body and increased blood flow. Other studies have shown that growth hormone increases the positive nitrogen balance that is needed for the formation of new tissue. Daily injections of growth hormone were particularly effective for increasing the nutritional status of patients after fractures, severe infection, or gastrointestinal surgery. In a double-blind trial, patients who were receiving IV feedings were also given injections of either growth hormone or a saline solution. The patients on growth hormone were able to conserve protein while the patients who received saline lost protein.

The healing properties of growth hormone are now under intense investigation. Although the most effective dosages and length of time that growth hormone should be given have still to be worked out, most researchers in the field believe that growth hormone has enormous potential for healing and restoring health after injury, illness, or surgery.

We have spoken about the incredible disease-fighting powers of growth hormone in this chapter. But in the next chapter, we will show how GH literally armors the body against illness by maintaining the shapeliness of youth.

Losing Fat, Gaining Muscle

Growth hormone may turn out to be the "magic pill" we've all been waiting for. It is unlike any weight-loss method that has ever been developed. With most regimens, you lose lean body mass along with fat, the biceps along with the belly. But your actual body shape changes very little. With growth hormone, on the other hand, many people do not lose weight on it and some actually gain weight. But they are incredibly pleased with the results. Why? Because they have resculpted their bodies, paring away the fat at the same time that they have increased their muscle mass. And because muscle weighs twice as much as fat, they may weigh more but are harder and leaner than ever. Growth hormone recontours the body. And, as we will see later in this chapter, a change in body composition means a change in body physiology. Your body not only looks younger, it functions younger!

WHAT HAPPENS TO OUR BODIES WITH AGE?

An aging body is like the creation of a Michelangelo sculpture, only in reverse—you move from the chiseled body of your youth back to a marble blob. Once we hit thirty, it's all downhill. The lean body mass (LBM) of all our organs starts to shrivel, while the adipose mass (AM), or fat mass, increases. Between the ages of thirty and seventy-five, the liver, kidneys, brain, and pancreas atrophy by 30 percent on average. In men between the ages of forty and eighty, the LBM declines about 5 percent per decade and in women by 2.5 percent per decade. At the same time the body fat in both sexes is expanding. By the late thirties, men are starting to gain fat in the gut. The upshot is that men between the ages of thirty and seventy don't gain weight, *but their lean mass shrinks by 30 percent and the fat expands by 50 percent.* Women as they age put fat on their hips. But when women start to go through menopause, the belly roll begins for them also.

These bodily changes are not just an affront to vanity, but a threat to health and longevity, according to gerontologists. First, the amount of aerobic power is directly connected to the amount of LBM. Second, the shrinkage of the vital organs means that they can't do their jobs as well, whether it is the heart pumping, the muscles lifting, or the kidneys clearing metabolic wastes from the blood. Third, as the abdominal fat rises, so does the risk for heart attack, hypertension, and diabetes.

It has become commonplace to blame our turning into blubber on a couch-potato lifestyle. But studies of exercising versus sedentary healthy men showed that only one-fourth of the atrophy of LBM can be blamed on disuse, while three-fourths is caused by the aging process. Daniel Rudman had a different idea. In 1985 he advanced his hypothesis that "the erosion of lean body mass and the expansion of adipose mass" was caused at least in part by the decline of growth hormone.

STUDIES WITH GH

By 1989, when the first long-term studies on the effect of growth hormone on adults were published, it was clear that Rudman was on the right track. In one study by St. Thomas' Hospital in London, of twenty-four adults with growth hormone deficiency, half the group was put on growth hormone and half the group was given a placebo. At the end of the six-month period, the hormone-treated group had no change in weight. But they had lost an average of 12.5 pounds of fat and gained an average of 12.1 pounds of lean body mass, which is mostly muscle. The increase in their lean body mass was 10.8 percent. Their bodies became sleeker and more tapered, as shown by a significant decrease in their waist-to-hip ratio. Although this study was conducted on patients with pituitary disease, the researchers note that "recombinant growth hormone reduced fat mass by about 20 percent, which suggests that growth hormone has a regulatory effect on fat mass in normal adults."

If these findings were limited to people whose pituitary glands had been removed or damaged by disease, they would be interesting but of limited significance. Only a few thousand people in the United States would be affected, but if the three-quarters of the aging changes that Rudman talked about were related to growth hormone deficiency and could be reversed by replacing it in the body, then it is not hyperbole to say that it would affect tens of millions of people and would change forever the way we think about aging.

Rudman's study showed that elderly men were just like young adults with pituitary disease. *In just six months of treatment, the men who were between the ages of sixty and eighty-one gained an average of 8.8 percent in lean body mass and lost an average of 14.4 percent in fat mass. In addition, the skin thickened by 7.1 percent, the bone density of the lumbar spine increased by 1.6 percent, and the liver and spleen grew by 19 percent and 17 percent, respectively.* In a follow-up study, he found that the same trends held after twelve months of treatment with GH, with the *lean body mass increasing by 6 percent, the fat decreasing by 15 percent, skin thickness up by 4 percent, with an 8 percent growth of liver and 23 percent of spleen.* The growth of the spleen is especially significant, Rudman pointed out, since the spleen, like the thymus, controls aspects of immune function that decline with age.

That the growth hormone was the engine for these changes could clearly be seen when Rudman stopped giving the hormone. The men's bodies fast-forwarded to old age. And if that weren't proof enough, consider the fate of the untreated controls. At the same time that the treated men were morphing into younger versions of themselves, the untreated control group was careening downhill, their lean body mass and organs shrinking by an average of 2.5 to 4.5 percent a year.

STUDIES OF OBESITY

The same shape-shifting is seen when patients are treated for obesity. In a 1994 short-term study of nine obese women, just five weeks of growth hormone treatment was enough to show a marked difference in body composition. This was a double-blind, placebo-controlled, crossover study, which means that all the women received both hormone treatment and placebo at different points. With the growth hormone treatment, the women lost an average of more than 4.6 pounds of body fat, a reduction of 5 percent of their total body mass. At the same time, their lean body mass increased by 6.6 pounds. Most of the fat that the women lost was in their abdominal area. In another study of people who were more than twice their ideal weight, growth hormone therapy also caused loss of fat regardless of their levels of growth hormone.

Two studies by Dr. David Clemmons, chief of the division of endocrinology at the University of North Carolina in Chapel Hill, found that growth hormone combined with dieting accelerated weight loss. These were double-blind placebo-controlled studies in which the patients received injections of either growth hormone or saline. All the people participating in the study were between 35 and 60 percent over their ideal body weight, making them obese but not morbidly obese, which is generally defined as 100 percent over ideal body weight. They were all put on a diet that cut their caloric intake by about 50 percent.

In the studies, which were for either six weeks or eleven weeks, growth hormone combined with diet caused a 25 percent acceleration in the rate of fat loss above and beyond the effects of diet alone. In the eleven-week period, the growth-hormone treated subjects lost thirty to thirty-two pounds, compared with twenty to twenty-five pounds in the controls. And while the controls lost muscle along with fat, the treated group held on to their body tone.

THE CENTRAL OBESITY FACTOR

That "little roll around the middle" is not only ugly, it represents a very real threat to one's health. There are two kinds of body fat, central fat and peripheral fat. The central fat around the waist and abdominal area is associated with a number of diseases, including atherosclerosis, type 2 diabetes, and syndrome X, which is characterized by high blood pressure, insulin resistance, unfavorable cholesterol profile, and premature atherosclerosis. Peripheral fat on the arms, legs, or hips is less related to health problems in general.

Americans are getting fatter. In the past 35 years we have gone from one-fourth of our citizens being chubby to one-third in 1980 to a *majority* in 1996. At the same time that we were pushing away butter and other fats, we were piling up on the pasta and other carbs. The result is that the rate of heart disease fell, but the incidence of diabetes rose. And the national bloating continues, with teenagers and even young children being fatter than ever. Much of the blame for this excess poundage can be placed on lifestyle factors, stuffing our faces while we channel-surf the cable. This creates a double whammy, since obesity and lack of exercise decrease the secretion of growth hormone, making it even harder to lose weight.

Fat men make 25 percent less growth hormone on a daily basis and have a pulsatile release of GH that is three times lower compared with normal-weight men. Rudman and other scientists have shown an inverse correlation between the amount of body fat and growth hormone secretion. The more body fat you have, the less growth hormone you release, and, conversely, the less body fat you have, the more growth hormone you have.

It now appears it is not just any fat that is the problem. In unpublished studies, Bengtsson and his group have found that it is the deep cushion of fat underlying the gut, the so-called central body fat, that lowers the amount of hormone that is secreted in growth hormone pulses. Blackman, who is looking at measurements of deep abdominal fat in his NIA growth hormone study, found the same thing. "Being apple-shaped rather than pear-shaped is not good for you not only because it is associated with an increased risk of diabetes and heart disease, but it also lowers your growth hormone," he says.

BUDGING CENTRAL FAT

For older people, trying to lose that central fat is like trying to move a block of concrete. This is where growth hormone shines. It is like metabolic liposuction, vacuuming out the fat from under the belly.

In six months of treatment, growth hormone–deficient adults at Sahlgrenska Hospital lost 20 percent of their body fat. And most of this fat loss was in the abdominal fat, which decreased by 30 percent, compared with a reduction of 13 percent in the peripheral area.

Just how effective GH was in getting rid of the central fat can be seen on a CT scan. After only twenty-six weeks of treatment, dramatic reduction can be seen in both the subcutaneous fat, just under the skin, and the visceral, or deeper, layers of fat. "We expected that the body composition would change with growth hormone treatment," says Bengtsson. "But we did not expect the magnitude of change. They were enormous. A 20 percent decrease in fat in six months, a redistribution of fat from abdominal depots to peripheral depots, and an increase in total body potassium and total body nitrogen which is the anabolic [body building] action of growth hormone."

MIGRATING BELLY BUTTONS

Forget stomach stapling. Belgian physician Thierry Hertoghe, who has made careful body measurements before and after two months' treatment with growth hormone and other hormones that decline with age, finds the results can rival that of a cosmetic surgeon. He typically records a shrinkage of about 23 to 30 percent in the love handles. It's not just the abdomen that shrinks, he says, "it is the upper side of the hand, the backside of the foot, the underside of the thigh gets thinner, while the upper side of the thigh gets thicker with more muscle." In fact, according to Hertoghe, the skin hanging from the upper arm of a woman signals the loss of growth hormone. When she has enough growth hormone, it tightens right back up.

The decreased fat and increased muscle tone makes for one of Hertoghe's most striking findings. The belly buttons of all his patients move up about 2.5 centimeters (almost one inch). "The belly buttons buy a beautiful ticket so they can travel," he says.

MECHANISM OF FAT LOSS

Growth hormone accelerates the burning of fat by making it available as fuel. It is like taking wood from the shed and putting it on the fire. By increasing the free fatty acids, growth hormone makes the fat stores available for energy production. In children who are growth hormone–deficient, injections of GH reduced both the size and amount of the fat cells.

Fat cells have growth hormone receptors, and when the growth hormone binds to the receptors, it triggers a series of enzymatic reactions in the cell to break down fat. This is called lipolysis. The hormone also gets rid of fat by increasing the energy expenditure overall so that you actually burn calories faster—a neat metabolic trick.

Another mechanism of fat loss, proposed by Rudman and his group, is that growth hormone counters the effect of another hormone, insulin. Whereas insulin promotes *lipogenesis*—the creation of fat cells—growth hormone fosters *lipolysis*—the destruction of fat cells. The way it works is this. Insulin is a kind of automatic garage door opener, letting glucose, amino acids, and fats enter the cells. Growth hormone blunts the receptivity of fat cells to insulin, in effect blocking the electric eye of the garage door opener. By constantly opposing the action of insulin, youthful levels of growth hormone keep the cells thin. Without growth hormone to inhibit the effect of insulin, the fat cells are free to expand. (In Chapter 17 we'll show you how you can change your diet to raise GH and lower insulin.)

UNLOCKING GROWTH HORMONE IN OBESITY

Fat has the same effect on the body as aging, reducing the production of growth hormone. But studies have shown the blockage of growth hormone due to obesity can be overcome. In Santiago, Spain, a group of researchers headed by Fernando Cordido used a combination of growth hormone–releasing hormone and a new growth hormone–releasing drug called GHRP 6 in obese patients who were more than 130 percent over their ideal body weight. The two compounds together had a synergistic effect, causing a massive release of growth hormone in these patients. They concluded: "The massive GH discharge in obese subjects after the combined administration of GHRP 6 and GHRH indicates that the blunted GH secretion in obesity is a functional and potentially reversible state."

This bears out the experience of many people who go on growth hormone and lose fat that they have been trying to get rid of for years. From Howard Turney, who trimmed nine inches off his waist, to Chein, who lost his "huge pot belly," to women like Sandy Bernstein, who report that they lose the unsightly puckers of cellulite, the fat comes off without people having to change their daily routine.

According to Dr. Edmund Chein, patients who are obese lose about 10 to 12 percent of body fat every six months on HGH therapy! "For instance, a person who is excessively overweight, say 5'4" and weighing 200 pounds, can lose 24 pounds of fat every six months. And it will continue until your fat percentage returns to normal." HGH is unlike any other method designed to take off fat, says Chein, because it works with your body rather than against it. When you go a diet, you almost always gain the weight right back again because you haven't changed the hormonal message that tells the body to store fat. Growth hormone changes the message, he says. It tells the body to get rid of the fat. "Hormonal replacement therapy is the only effective weight-loss regimen in which we don't tell people to count calories or stop eating tasty things."

BEEFING UP THE BODY

At the same time that patients are losing weight with growth hormone, they are putting on muscle mass at the rate of about 8 to 10 percent every six months, according to Chein. "With our patients we don't talk about fat, he says, "we talk about percentage of fat and percentage of lean muscle. You are reversing, losing fat and gaining lean muscle. Your net weight may stay the same or even go up."

The chemical reactions that give rise to the synthesis of new proteins, cells, and tissues is known as anabolism. Growth hormone working through IGF–1 promotes the uptake of amino acids (the building blocks of protein) in the cell and enhances the synthesis of DNA, RNA, proteins, extracellular proteins, carbohydrates, and sugars. The result is an increase in cell size and rate of cell division.

Growth hormone also builds muscle by conserving nitrogen in the body. Ordinarily, when people lose weight, they excrete nitrogen, which is needed for muscle tissue. But growth hormone keeps the body from losing nitrogen. Second, it increases protein synthesis without increasing the breakdown of proteins in the cell (see Chapter 5). The net effect is a gain in lean body mass. Growth hormone also stimulates muscle growth directly by increasing the number of muscle cells.

Studies in London, Denmark, and Sweden of growth hormone–deficient adults all show that hormone replacement increased muscle mass and decreased fat mass. A long-term follow-up of the

Isometric Strength in the Quadriceps

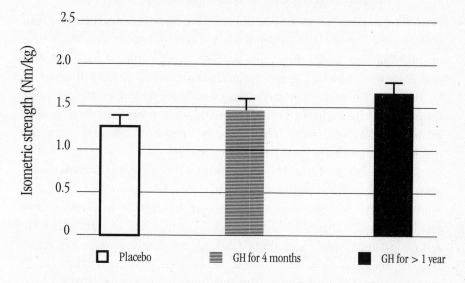

Isometric strength in the quadriceps muscle increased significantly after 16 months of GH treatment. Data taken from: Jørgensen J.O.L. et al., *Acta Endocrinol* 125 (1991): 449–53.

Danish study revealed that after one year the muscle-to-fat ratio was even better. Not surprisingly, the bigger muscles translate into greater strength, with continuing improvement over time. Two studies showed that patients who had been on GH therapy for a year had better muscle function compared with those who had been treated for only four months.

The gain in lean body mass in older people means that vital organs that have shrunk with age start growing again. The increase in breathing capacity seen in patients with chronic obstructive lung disease is most likely due to actual increases in lung size. (Dr. Baxas found that patients on HGH can hold their breath and stay under water longer.) The growth in heart tissue makes for stronger hearts, a greater ability to contract, a stronger pump to push the blood through the circulation. An increase in kidney size means a greater ability to filter the blood. Accelerated cell division means speeded-up healing and repair after injury. And an increase in muscle mass translates into greater strength and endurance and vastly improved quality of life. Denser bones mean less chance of a hip or leg fracture that can change a functional older person into a crippled, dependent one. The growth effects of growth hormone may even extend to the eyes and the brain (see Chapter 12).

Just imagine what these changes in body composition might mean to the aging population in the United States. In a recent article in the *New York Times* magazine section, old people in their eighties and nineties found little solace in their survival. They felt crushed by the burden of their years. "Successful aging" was a cruel joke, they said. Life held no more meaning for them. Here is where growth hormone therapy may have the most visible and dramatic impact, on the oldest members of our society.

Growth hormone stimulation with GH-releasing supplements after age thirty-five could radically improve the picture of advanced aging so that we can live out our lives in dignity and autonomy until the end. Growth hormone replacement for the frail elderly whose pituitary has essentially burned out might restore the function of these people to what it was twenty years earlier. With increased brawn and bone, doddering oldsters could throw away their canes and walkers, lift packages without asking for help from others, prepare their own meals. The effects of HGH on the brain would improve their memory, increase their thinking power, encourage them to socialize, and, in general, lift their spirits. The nursing homes would start emptying out, and the very old might find the meaning of life once again as people have found it from time immemorial—in the living of it.

GROWTH HORMONE AND INSULIN

Until quite recently, it appeared that a major drawback to the use of growth hormone was its effect on insulin. It appeared to counter the effects of insulin on the metabolism of glucose, making people more insulin resistant. For this reason, doctors excluded anyone who was diabetic or prediabetic from using the drug. Rudman and others had found that some people had increased blood sugar, a sign of insulin resistance, and there was concern that long-term administration of growth hormone could make some people diabetic.

In his earlier studies, Bengtsson found that growth hormone treatment increased insulin resistance in patients after six weeks of treatment. But after six months of treatment, their insulin sensitivity returned to where it was before starting treatment. "If you look at the large studies with growth hormone," he says, "there is no increased incidence of diabetes."

Bengtsson believes that the profound changes in body composition after six months may have counteracted its effect on insulin. There is a close association between central obesity and insulin resistance. By getting rid of abdominal fat, you can induce greater insulin sensitivity. If insulin is more effective in disposing of glucose, then the glucose does not remain in the cells where it can cause protein-glucose complexes (AGE—advanced glycosolation endproducts), a form of cellular garbage that accumulates with age and interferes with the workings of the cells. While it is yet to be proven, it seems reasonable to assume that over the long run, stimulation of growth hormone by getting rid of abdominal fat could prevent type 2 diabetes or even reverse the disease process. "Growth hormone promotes the action of insulin," says Hertoghe. "When we use GH, it seems to direct the action of insulin towards putting sugar into the cardiac, muscle, and nerve cells, rather than into fat cells. It seems that growth hormone may help diabetes."

BODY COMPOSITION AND LONGEVITY

If, as Freud said, anatomy is destiny, your body shape determines your longevity. While you may not be able to change your anatomy (at least without major surgery), you can control your shape. And by changing your body composition, you can lengthen your life.

As we discussed, obesity, especially fat stores around the middle, are associated with degenerative diseases like diabetes, hypertension, and cardiovascular disease. In a famous long-term study of Harvard male alumni, the researcher divided the graduates into five groups based on their body mass. Those who fell in the heaviest group had shorter than average life spans, while those who were the leanest lived longer than average.

Remember the life span studies that showed that animals whose caloric intake was reduced 30 to 50 percent far outlived the control groups that were allowed to pig out? The diet-restricted animals are leaner, have more muscle mass, and have increased protein synthesis. All these changes are exactly the ones seen with growth hormone replacement therapy.

By reshaping your body, you will live longer, healthier, and sexier. And you will look years younger, since growth hormone also does wonders for your face, as you will see in the next chapter. In the second section of this book, we will show you how a total program of diet, exercise, and growth hormone releasers can give you the body shape you have lost or always longed for.

Losing Wrinkles, Growing Hair

Growth hormone is the only anti-aging treatment known that actually makes people *look* younger. Even creams and lotions that contain antioxidants like vitamins E, A, or C, retinoic acid, or fruit acids which have been shown to erase fine lines do not stop the skin from sagging and sinking. Proper diet can help the complexion, aerobic and weight-training exercise can give you a lean, muscular body, antioxidants can reduce the free-radical reactions that play a major role in causing disease, even hormone treatments like estrogen and testosterone can restore many body functions. But only growth hormone therapy can take a decade or more off your face.

THE CAUSES OF SKIN AGING

Like the movement of the tectonic plates that leads to fissures in the earth, so do the shifting proteins under the surface lead to the disaster that is skin aging. The key proteins of the skin are collagen and elastin, which form an underlying three-dimensional matrix that supports the skin like the armature of a clay sculpture. But over the years, there is a decrease in the manufacture of new collagen and elastin as well as change in their structure, which causes the matrix to loses its neat, intertwined organization. This is when cracks and wrinkles appear.

At the same time that you are gaining fat in your belly, you are losing the fat in your face. According to Hertoghe, there is a look about the face that people have who have growth hormone deficiency syndrome. These are people in their fifties and sixties who have saggy faces due to loss of fatty facial tissue of 1 to 2 percent per year. They are losing the cushion of fat under the skin as well as body water. The thinning of the skin, the sagging quality, the overall droop of the facial features, the loss of subcutaneous fat, and decrease in water turns the ripe plums of young faces into the prunes we associate with aging.

LIFTING THE FACE WITH GH

People on HGH start looking visibly younger usually within a few weeks of treatment. Often the first person to notice the difference is the spouse or significant other. One man said his wife looked at him in astonishment after the second week of treatment and said, "Your crow's-feet have disappeared." Not only do the fine lines vanish and deeper wrinkles recede, but the face actually undergoes a change of contour in the same way that the body composition changes. The fat decreases and the muscles increase, so that puffs of fat under the eyes evaporate, while the facial muscles that lift and hold the skin become stronger. And it increases the synthesis of new proteins, helping to restore the underlying matrix.

John Cantwell, M.D., of San Jose, California, who recently started using growth hormone in his medical practice, added HGH to the regimen of two fifty-year-old men who were already on hormone replacement therapies. Within three months, the men had fewer wrinkles, and their skin texture and elasticity was much improved. "They looked so much younger and felt so wonderful, they both decided they didn't need to come in for further checkups," says Cantwell.

On an NBC special on growth hormone, Tom Brokaw remarked on how young the people featured on his show looked. William B. William, shown at age eighty galloping across a polo field, looked about fifty-five, said Brokaw. Paulene McNair, then fifty-two, looked "maybe forty." Indeed, seeing them in close-up, it was almost impossible to believe that they were not the ages that Brokaw said they looked. It is not just their faces, which were unusually smooth and unlined, but the vitality and enthusiasm that shone from within. In a shot of McNair talking with women half her age, whom she said she used to envy, she seemed like their contemporary. And their bodies were as youthful as their faces. The shoulders and arms and legs of McNair shown when she was wearing a bathing suit and a close-up of William's thigh as he gave himself an injection revealed the same remarkably age-free skin.

HORMONAL "PLASTIC SURGERY"

Growth hormone is as close as you can get to having a face-lift without going under the knife. Since one of its main effects is stimulation of protein synthesis, growth hormone and its co-factor, IGF–1, activate the production of skin proteins like collagen and elastin. In animal experiments, it increased the strength and collagen content of the skin. Collagen and elastin are the foundation of the epidermis. The turgor, or bounciness, that is characteristic of young skin also returns (more about this below). "If you give someone growth hormone," says Chein, "you improve the collagen synthesis and increase the muscle mass underneath the wrinkles. You are really doing medical cosmetic plastic surgery."

According to Vincent Giampapa, M.D., a plastic surgeon and pioneer in anti-aging research, "As you age, the skin gets that loose draggy look and a lot of fine crepey lines. With growth hormone, the elastin and collagen subjectively are much better, and the fine crepey lines seem to be much improved. The skin turgor is markedly better. The skin doesn't seem to hang that much."

THE PINCH TEST

A rough measure of skin turgor is the pinch test, where you pinch the skin on the back of your hand and see how long it takes to return to its flat position. This is one of the self-test biomarkers of aging (see Chapter 13) since the ability of your skin to snap back is an indication of the elasticity which goes down with age. Chein is fond of stopping complete strangers and asking them if they will try a small age test—pinching the skin on the back of their hands to see how fast it returns. Before he started on HGH, says Chein, the speed at which his skin returned to a flat position was like that of other people in their forties. One year after being on the drug, he was beating people in their thirties. Now, he says, he is in competition with people in their late twenties.

One seventy-three-year-old man we interviewed for this book had the "snap back" of a young man. Although he is a carpenter and uses his hands all the time, the growth hormone treatments have thickened and toughened his skin, so it is "less likely to be banged up," he says. Most remarkable was the appearance of his hands themselves—firm, full, strong, unlined, with no age spots, the hands of a much younger man.

Elmer Cranton, M.D., director of the Mount Rainier Clinic, in Yelm, Washington, who has been treating people with growth hormone for the past five years, has also noticed reversal of skin wrinkling to some degree. "The skin gets thicker from the protein build-up," he says. "And the dehydration that occurs with age is reversed almost immediately."

Growth hormone also has tissue repair effects that may take several months to become visible. One patient who has been on growth hormone therapy for two years noticed that the sun damage on his shoulders and arms from years of living in the tropics was clearing up. "I feel that the texture is better," he says.

Even the hands and feet look younger. The brown age spots fade, in some cases disappearing altogether. The veins become less prominent and there is the same increase in moisture and skin turgor. This means that unlike a face-lift or chemical peel, the hands, feet, indeed all the skin of the body, experience age reversal.

REGAINING BODY WATER

There is nothing quite like the feel of a baby's skin. It is soft, smooth, plump, with a kind of delicious squishiness. That's because a baby is mostly fluid, with water making up about 90 percent of its body. By adulthood, the body water is down to 60 percent, and by the time we reach old age, it is a mere 40 percent of our total body mass. In a very real way, aging is a drying out process from the juiciness of youth to the shriveled, desiccated, puckered face of old age. You can see the same dehydration in young adult patients with GH deficiency due to pituitary problems. Their skin is dry and thin and many of them look old for their age.

Growth hormone turns the prune back into a plum. This is

through a little understood effect called antinatriuretic action, which acts through the kidneys to retain sodium and water. It is this effect that is also responsible for the most commonly reported problems with the use of growth hormone—swelling, pain in the joints, and carpal tunnel syndrome. All these effects can be eliminated by lowering the dosage. As clinicians have gained experience with using GH, they have learned to individualize the dosage so that the they retain the desired benefits without the adverse side effects.

HAIR GROWTH

GH appears to have a tonic effect on hair. While there is no scientific data, patients report improvement in the quality of the hair. In the self-assessment performed by the 202 patients in the study by Terry and Chein, an astounding 38 percent reported new hair growth (see Chapter 3). "It grows faster and texture is thicker," says Elmer Cranton, who says his own gray hair has turned brown. Other patients also find that their hair is regaining its original color. Some people, like forty-three-year-old James Phelts, claim that new hair is sprouting in previously bald areas. One of Dr. Baxas's patients, Geoffrey Clarkson, also reports increased growth of new hair on his scalp. Dr. Giampapa's patients attest to thicker hair. "It's real," says Dr. Lord Lee-Benner, "there is hair growing on people's heads." Fingernails and toenails also grow faster and look better, according to McNair, who says that even her eyelashes got thicker.

GROWTH FACTORS AND SKIN MORTALITY

Growth factors are growth hormone–like proteins and peptides (small proteins) that promote the growth, nourishment, and repair of cells. Growth hormone helps stimulate the production of these factors, which act like a kind of rescue squad when the tissue is damaged by injury, illness, or aging. "Growth hormone is the general and the growth factors are the foot soldiers that do the work locally in the tissue," explains Eric Dupont, Ph.D., head of Aeterna Laboratories in Quebec. We have spoken here a lot about IGF–1 (insulinlike growth factor 1), but this is just one of many growth factors (see Chapter 21). A number of these factors, such as epidermal growth factor (EGF), work to repair and maintain the architecture of the skin.

Surprisingly, the skin is both mortal and immortal. The top layer of the skin, the epidermis, is immortal. Retinyls, like retinoic acid (Retin-A), and alpha hydroxy acids (AHAs) stimulate this immortal layer and can be used on a daily basis to exfoliate the skin at any age. The epidermis can produce thousands of generations of cells as needed. The dermis, which lies under the epidermis, is mortal. Its cells divide about one hundred times after conception. Growth factors work on the mortal layer of the skin. As you age, the dermis gets thinner, loses elasticity, and becomes less and less responsive to the growth factors and cytokines (cell messenger peptides) that signal the skin to divide. This is the point at which growth hormone and a "cocktail" of growth factors may stimulate the old cells to function like younger ones. Recent research shows that a growth factor called FGF (fibroblast growth factor) can stimulate even aging fibroblasts to produce the matrix proteins of collagen and elastin. And they can do it in concentrations that are close to those of cells from younger age groups. The use of growth hormone and growth factors offers tremendous short-term promise in their ability to remake our skin and face. (See Chapter 21 for a full discussion of the various growth factors now available or in development.) The era of the biochemical bioengineered face-lift is upon us.

GROWTH FACTOR COSMETICS

The most exciting new skin creams and lotions use a combination of retinoic acids like Retin-A, alpha hydroxy acids (AHA), antioxidants like vitamin C, and growth factors. One such regimen is Agera, which combines technologies to produce dramatic results, especially with sun-damaged skin. Recently the researchers at Biosyn, which makes Agera, have formulated cosmetics that use peptides and a highly stable form of vitamin C to treat the discoloration associated with aging and sun overexposure.

Vincent Giampapa, M.D., is now researching a number of growth factors and other compounds at the National Skin Institute in Montclair, New Jersey. "Basically, we are giving the cells of the skin, the fibroblasts, what they need to repair their DNA faster and make more collagen and elastin," he says. The agents they are working with include RNA, which helps restore skin turgor; hyaluronic acid, which helps retain moisture and regenerate collagen; yeast cell wall extracts; and retinyl palmitate, which is like vitamin C with vitamin A hooked to it, to promote faster cell division. The RNA comes from amino acid and nucleic acid supplements such as Spiralina and AFA algae, both of which are available from health food stores. The former is a saltwater blue-green algae, while the latter is a freshwater blue-green algae. "These are the richest sources of nucleic acids and essential amino acids in a predigested form, which means that as soon as you swallow them, they're in the blood-stream," he says. He also uses growth hormone releasers like arginine, ornithine, lysine, and glutamine (see Chapter 16).

Growth hormone is the master ingredient in this recipe, he says, since it speeds the entrance of the amino acids and nucleic acids into cells, where they can be used for cell renewal and restoration. With this regimen, Giampapa contends, "In about a month a half, you can see visible improvement. The skin looks plumper and less lined. It also helps with skin color, making the skin look less mottled, increases the microcapillary system so that you get a better blood supply to the face, and reverses the damage that comes from a life time of abuse from the sun, poor nutrition, smoking, and bad diet. These things help reset the aging clock of the skin at slower speed. So, for instance, instead of your skin aging at 60 miles a hour, with each month that goes by, it slows up that it is now running at 50, 40, 30, 20 and will stay there."

He also uses a similar mix of nutritional supplements and growth factor releasers for a minimum of two weeks in people who are undergoing cosmetic surgery. This greatly enhances the repair capability of the skin and cuts down the healing time, he says.

Beyond the cosmetic effects are the psychological ones. Growth hormone, as we will see in next chapter, acts to heighten your sense of aliveness and well-being. You literally stop feeling your years. Instead of a down, deadbeat look that many people get as they age, the faces of people on growth hormone stimulation shine with youthful enthusiasm. And because growth hormone works to stimulate the gonadal hormones that drive the libido, the face radiates a kind of sexuality you may have thought disappeared from your life years ago.

George Bernard Shaw said, "Youth is wasted on the young." Well, not anymore!

GH and Sex:
The Hard Facts

Sexuality is part and parcel of our being. When we are young, it seems to rule our lives. Whether we are men or women, we are driven by the need to look good, smell good, dress well, make ourselves attractive to those we wish to attract. Advertisers know this and play upon our conscious and unconscious desires, to sell everything from soap to automobiles. To be sexy seems the very essence of youth. And if the time comes when we no longer feel the need to attract and be attractive to others, it seems as if something has gone out of our lives forever. We have aged in a very real sense.

CAUSES OF DECLINE IN SEXUAL FUNCTION

To understand how growth hormone can revive a flagging sex life, let's look at some of the root causes of male impotence. While it used to be thought that psychological factors were the primary reason for loss of virility, sex experts now believe that physical causes are responsible in 75 percent of cases, and in 90 percent of cases in men over fifty. According to a National Institutes of Health Consensus Panel on Impotence, nearly half the incidence of erectile dysfunction in men over fifty is caused by arteriosclerosis of the penile arteries. The same buildup of cholesterol plaques that narrow the coronary arteries and can cause a heart attack occurs in the tiny blood channels that feed the penis. The swelling of the erect penis is a direct effect of the engorgement of blood. In fact the failure to have erections can be like the death of a canary in a coal mine—a forewarning of danger. According to an article in *Postgraduate Medicine* by J.E. Morley, erectile dysfunction has been shown to be a preceding sign of a heart attack or stroke. The same risk factors that contribute to heart disease, namely a poor lipid profile, including high total cholesterol, low HDL, high LDL, high triglycerides, and cigarette smoking are implicated in impotence. Other common diseases that are associated with sexual dysfunction are insulin-dependent diabetes and untreated hypertension—unfortunately, the drugs used to treat hypertension can also cause impotence.

Another major cause of erection failure is decreased testosterone levels. One of the best kept secrets is that men go through a male form of menopause called andropause. While menopause occurs in all women and is accompanied by an almost complete shutdown of ovarian estrogen, the male version is not universal and the decline in testosterone is much more gradual. The average healthy man in his late forties or early fifties has one-third to one-half the level of testosterone that he had in his twenties. An even greater drop occurs in unhealthy men. "The male menopause is a real phenomenon," says Blackman, "and it does similar things to men as menopause does to women, although less commonly and to a lesser extent." Some men, for instance, experience hot flashes, sweating, aches and pains, loss of energy, dry skin, and depression. The decline of testosterone also contributes to the same signs of aging as a decline in growth hormone does, such as increased body fat, decreased muscle mass, loss of libido, and failure to have erections. Sometimes the loss of testosterone is due to pituitary disease. Another endocrine problem is underproduction of thyroid hormones, which can lead to sexual dysfunction. (For a fuller discussion of hormonal replacement with testosterone and thyroid hormones, see Chapter 15.)

Of course lifestyle factors play a role, particularly heavy drinking and cigarette smoking. In the past few years smoking has become recognized as a major culprit in cutting off the circulation of the penile artery. If you're a smoker, giving up cigarettes to save your sex life may be a stronger motivator than merely saving your life.

THE EFFECT OF GH ON MALE SEXUALITY

The decline of male potency parallels the decline of growth hormone release in the body. At puberty, when growth hormone is at its peak, a boy is in an almost constant state of arousal. By young manhood, the time between erections is already decreasing. At this point, he is usually able to have sex every day and maybe several times a day. By his forties, he may be down to two or three times a week, and by age sixty to about once a week or less. At the same time, the rate of impotence increases. Every study of male sexual function shows a decline with age. According to the classic survey by Kinsey, under the age of forty about 2 percent of men are impotent. By the age of eighty, 75 percent are incapable of having or sustaining erections. The Massachusetts Male Aging Study (MMAS), in a random sampling of 1,290 men, found that complete erectile dysfunction increased from 5 percent at age forty to 15 percent at age seventy-nine, with difficulty in having erections experienced by 52 percent in the older age group. Is the decline in growth hormone and sexual performance a coincidence, or is there a cause-and-effect relationship between the two? In the first clinical analysis of the effects of HGH replacement in 202 people, including 172 men, by L. Cass Terry and Edmund Chein, fully three-quarters of the participants said that they had an increase in sexual potency or frequency. And 62 percent of the men reported that they were able to maintain an erection for a longer period of time. Since the vast majority on growth hormone therapy report an increase in sexual appetite and performance, we strongly suspect that growth hormone plays a pivotal role in sexual health.

GH AND FEMALE SEXUALITY

Women do not have a decline in sexual function that is comparable to that of men. A healthy woman can enjoy intense orgasms into old age. But a number of factors associated with aging can affect a woman's sexual pleasure. Starting around age forty-five with the loss of estrogen, and made progressively worse by the menopause, women commonly experience dryness and atrophy of the vaginal tissues. This can cause discomfort that can lead many women to avoid intercourse or have it less frequently. Lack of regular intercourse can then interfere with a woman's natural enjoyment and ability to have orgasm. In the same way as a man, she has to use it or risk losing it.

Many older women also report a lack of libido. If they have a partner, they may wish to have sex less frequently. If they are alone, they may contend that they prefer it that way, that they have lost interest in men and sex. This too is often related to hormonal decline, not only of estrogen, but of testosterone, DHEA, and growth hormone. Estrogen replacement can relubricate the vagina, while testosterone and DHEA can revive flagging sexual interest. (See Chapter 15 for a full discussion of the benefits of total hormonal replacement in women.)

But the most powerful effects come from growth hormone, which can reawaken the sex drive and orgasmic delight that a woman reveled in in her youth. Women on GH therapy report increased libido, heightened pleasure, and the equivalent of greater potency in men in the form of multiple orgasms. Part of the women's heightened sexuality comes from their sense of being more attractive, with firmer skin and well-toned bodies. There is nothing like looking great to put a person, male or female, in the mood for love. And the increased sexual activity for her, as well as for him, has a rejuvenating effect on the pituitary and the neuroendocrine axis (see chart on page 43 in Chapter 4). Growth hormone has the added benefit in women of alleviating menstrual and postmenopausal symptoms. For instance, Dr. John Baron of the Baron Clinic in Cleveland treated a thirty-five-year-old woman with severe premenstrual tension with growth hormone and progesterone. Her symptoms abated to the point where she is now scarcely aware of having her period. And a fifty-five-year-old female doctor had her hot flashes completely disappear with GH replacement therapy. Her FSH (follicle-stimulating hormone) levels also dropped from 60, which is menopausal, to 8, which is normal for a young menstruating woman.

Although there are no clinical studies that specifically look at growth hormone and its effect on sexual function, the evidence is that growth hormone is pivotal in both the desire to have sex and the ability to carry it off. First, people who are growth hormone–deficient due to pituitary disease have decreased libido and sexual function. Second, when they are given growth hormone, according to Bengtsson, they report increased sexual drive and function as shown on their responses to the Nottingham Health Profile, a self-rating questionnaire that measures quality of life. Third, a number of people who are using growth hormone as part of an anti-aging regimen have offered testimonials.

BEARING WITNESS

According to all the physicians who use growth hormone in their practice, the sexual changes are striking, affecting both men and women. Both sexes report increased libido, while male patients say that they have better erections, better performance, longer duration, and decreased recovery time between orgasms.

The effect of growth hormone on the libido of both men and women is so strong, says Vincent Giampapa, M.D., that if both the husband and the wife are not on it, the marriage is in trouble. Sam Baxas agrees. "We find that the men start looking at young girls again, so we have to treat the wives also." He gave his own wife growth hormone along with estrogen and progesterone to relieve symptoms of menopause. As a result of the female hormones, "she is menstruating like a 20-year-old and behaves like a young girl. She's 65 and I'm 61, but we feel like a couple of young kids," says Baxas, who is also on growth hormone.

One patient who is feeling the hormonal imbalance in his marriage is a forty-nine-year-old business executive who has been married for twenty-five years. Their sex life had fallen into a comfortable rut in which they had relations once or twice a week, usually on the weekend, says the man, whom we'll call Fred. An adventurer and risk-taker whose work has taken him all over the world, he began using growth hormone after reading about the Rudman study. About a month after starting, Fred says, "I began feeling horny all the time. Suddenly I wanted sex almost every single day but my wife didn't feel the same way. I've never cheated on her, but for the first time I started looking around." He urged his wife to take the drug, but she refused, saying she didn't think she needed it. He is now thinking about leaving her. "Maybe it's crazy," he says, "But I feel like we're on two different tracks. I'm getting younger and she's just staying the same or getting older. It's a real problem."

It's not just the men who feel that the gonadal batteries being recharged. Dr. Elmer Cranton relates the story of a woman patient who began experiencing strong erotic urges after going on growth hormone. Repeatedly she urged her husband to take HGH so that they would both be on the same sexual wavelength. Eventually she gave up and left him for a younger man.

For some people, growth hormone treatment has meant a sexual awakening. A sixty-three-year-old divorced doctor, who hadn't been with a woman in five years, felt he was no longer interested in sex. "To tell the truth," he says, "I felt kind of like I was a neuter." After four weeks on HGH, he felt a surge of sexual energy. He started dating again and now has sex on an average of twice a week. A convert to HGH, he is now planning to use it in his medical practice.

One of Dr. Cranton's patients is an eighty-year-old man with severe congestive heart failure. Because of his advanced age, his doctors had ruled out a coronary bypass or heart transplant. Thinking that growth hormone might help his heart, Cranton recommended injections of HGH. In a few months, the man had recovered both his heart and sexual function. "His wife says, 'he's getting amorous again,'" chuckles Cranton.

With growth hormone, it seems you're never too old. Dr. John Baron at age eighty-two, who has been taking growth hormone for over a year, says, "I surprised myself. I'd say my libido and sexual function is as good as it was when I was 25. My affinity for the opposite sex is greater and I appreciate them more. It's made my life happier."

McNair tells of trying to call her eighty-year-old father one day over a period of several hours. After six months on growth hormone he had acquired a girlfriend and was driving a new Trans Am. As she tells the story, "When I finally got hold of him, I asked him where he had been and he said, 'none of your business.' I said, 'daddy, are you getting laid?' And he just roared. He said, 'You're my daughter and I don't have to tell you that.' But I knew perfectly well that was what he was doing,'" says McNair.

Lee-Benner has also noted a tremendous rejuvenation among his clientele. "I've had patients who are in their 80s tell me that hormone replacement therapy has returned them to a level that they haven't seen in fifty years," he says. Among his elderly patients are an eighty-four-year-old woman and her eighty-year-old husband. "They are a respected, successful retired couple," he says "Recently I got a call from him at six in the morning. He said, 'Doc, Hazel needs a man here. I need you to come over and give me a shot.'"

Some people might argue that these dramatic changes in sexual behavior are a placebo effect. As a physician, I am well aware that if someone believed that green M&Ms were an aphrodisiac, a handful of the candy might be enough to goose the gonads. But it seems almost impossible that a placebo could make a man interested in having sex after five years of feeling no desire at all or increase his activity from once a week to several times a day. Furthermore, the beneficial effects of a placebo fall off after a few weeks, while most of these patients have been taking growth hormone for at least six months and some for as long as five years, with most reporting an increase rather than a decrease in sexual activity. I dare you to take away their HGH now!

INCREASE IN ORGAN SIZE

It is well documented that growth hormone reverses the shrinkage of the heart, liver, spleen, and other vital organs. Now it appears that it may regrow the sexual organs as well. Starting in the fourth decade, the penis and the clitoris actually start to shrink. But many patients report that with growth hormone the penis returns to its youthful dimensions. The same reversal occurs with the clitoris in women. According to Chein, "when you restore the hormones to that of a 20-year-old, the man's penis and the woman's clitoris will return to their original size. This has not been reported in the journals yet, but these are the results from my patients after six months, one year, two years, and longer." Baxas, whose has been using growth hormone in his practice since 1990, says, "The penis definitely gets bigger. It's an organ that can increase in size. My patients tell me that they are more aware of sex, more sexually active. They go back to having multiple orgasms. I've had men tell me, 'it is better now than when I was 20."

While acknowledging that growth hormone can boost penis size, Ahlschier cautions that "this is not something you can count on. It depends on such factors as having the correct amount of hormone in your body, whether you have good muscle tone, and your blood supply." Here is where the exercise effect of growth hormone is most valuable (see Chapter 18).

SEXUAL REJUVENATION

Along with the increase in size comes an increase in function. It would be hard to credit the claims of sexual prowess by users of growth hormone if they weren't so consistent. To a man, they report that the sexual falling off with age is reversed to what it was in their twenties or thirties. At age sixty-two, Lee-Benner says he can have sex every day, or even two or three times a day when he wants. "I'm not what I was when I was 20," he says. "And I still need hormone stimulation for optimal sexual function. But I find that growth hormone has intensified my orgasms back to what they were in my teens. I didn't know that my orgasms needed intensifying, but now that it's back I sure can tell the difference."

Often it's the partner who notices the difference. L. Cass Terry, the fifty-five-year-old Medical College of Wisconsin neurologist, says, laughing, "It scared my wife. She said, 'what's that doctor doing to you?'"

A fifty-year-old marine biologist, who wished to remain anonymous, says, "After age 45, things started to slow down. I don't know whether it was mental or physical, but my erections were not as firm or as reliable. I was used to having sex five or six times a week, on good days, making it two or three times. All that fell off to maybe once or twice a week. That is probably average. But it is not what I wanted. With growth hormone, I am probably not where I was when I was 25, but I'm back to what I have time for. And a couple of times a day is not straining anything."

HGH made fifty-five-year-old Allan Ahlschier, M.D., into a sexual dynamo. "I don't know whether it is psychological or not," he says. "I was recently divorced after 33 years. My sexual activity was once a week, which is not too far from average for my age." Now remarried, he reports having sex two to four times a day. "Once in the morning and once at night. I live close enough to work so that I can come home for some afternoon delight."

He believes that his bodybuilding program, particularly the leg squats and leg presses, are especially good for enhancing sexual prowess. The blood vessels that feed the leg and thigh muscles go right to the penile artery. This may be one of the reasons that his own penis has enlarged, he says. "I do know that the day I do my leg presses, my wife's in real trouble." Ahlschier points out that the average recovery time for fifty-five-year-old men after orgasm is eighteen hours. "Mine is like two minutes or less. Sometimes it doesn't ever go down!"

RESTORING SEXUAL VIGOR

Growth hormone reverses the age-related decline of sexual vigor and performance. And it does this at every level of bodily organization, from the cells to the organs to whole systems. At the cellular level, it increases protein synthesis and cell division involved in regrowth of organs. On the tissue level, the anabolic effects of nitrogen retention build muscle mass.

On the hormonal level, there is a direct feedback between the pituitary hormones and the gonadal hormones of the ovaries and testes. In many men, taking HGH is enough to bring testosterone levels up to normal. It is also interesting to note, says Karlis Ullis, M.D., that growth hormone plays a role in both male and female infertility, improving the effect of the reproductive hormones in the production of sperm and eggs. "This shows one, there is an enhancement factor and, two, it is rejuvenating. It also crosses into the brain and seems to activate the whole nervous system. People report more pleasure, which could be because it sensitizes the nerve endings."

On the systemic level, growth hormone improves cardiac function and circulation of the blood to every part of the body. When more blood is pumped into the penis, erections are firmer and can be sustained longer. Growth hormone has also been shown to increase the HDL cholesterol. In one study, raising the level of HDL cholesterol increased the likelihood of an erection, perhaps by increasing peripheral circulation via the plaque-removing properties of HDL. Growth hormone also accelerates the repair of tiny tears in the penis and vessel damages that come from trauma during intercourse, from riding bicycles, or from other activity.

EFFECT OF WEIGHT TRAINING ON SEX

One of the biggest benefits of growth hormone is that it increases your ability to exercise longer and harder (see Chapter 18). In combination with weight-training, especially leg presses and squats, it can increase the blood flow to the penile arteries. "The aging penis is like a leaky valve," says Ahlschier. "The blood is pumped in but if the valves are working, it gets pumped right out and there is no erection. If you look at the venous tone of body builders, you'll see huge arm and leg veins. Body building exercises increase the venous tone," he says. "There has been some controversy about whether body builders have larger penises. Some people contend that they have small penises. Nothing could be further from the truth. The penis is another muscle. If you are exercising and in good general health, all the system of the body are going to work better."

THE PRIMARY SEX ORGAN

The most important sexual organ lies not below the waist but above the neck. How we think and feel profoundly influences sexual functioning in both men and women. As we will see in the next chapter, growth hormone has a direct effect on the biochemistry of the brain, which, in turn, controls cognition, emotion, and mood. It also raises the cellular metabolism and energy level. People on growth hormone often say that they can't tell whether their enhanced sexual performance is due to psychological or physical factors. In truth the answer is both. Mind and body are intertwined. The neurotransmitters, which are the chemical messengers in the brain, have receptors all over the body. Sexual responsiveness is to a large degree determined by what happens with these neurotransmitters. When we are anxious or stressed out, it is difficult to get turned on. When we are relaxed, when we feel good, and when our juices are flowing, we become open to the sensual world around us, the sights, the sounds, the smells, the touch of lips, tongue, fingertips that delight and tease the mind. The pleasure center lights up and desire and arousal flow naturally.

Powering the Brain

What's the point of looking young, having a great body and the sex drive of a teenager, if your mind has gone south? Here is where growth hormone may have its most compelling impact, improving all those things that, when you get right down to it, people care about most: happiness, energy level, sleep, ability to think clearly and remember. Some people even see and hear better. But growth hormone may do more than reverse the mental and emotional decline that accompanies aging; it may prevent or help the most terrifying aspects of growing old, the mind-destroying diseases of Parkinson's and Alzheimer's.

THE EFFECT OF GH ON THE BRAIN

Although growth hormone is made in the brain, it is only recently that scientists have discovered that it has a powerful effect on the brain as well as the body. Some researchers believe that the fall in growth hormone levels with age may account for the shrinking of the brain with age and that giving HGH may regrow the brain. It also appears to protect brain cells from injury and oxygen deprivation, such as that which occurs with stroke. And a groundbreaking discovery by Bengtsson and his group in Sweden shows why growth hormone may be nature's Prozac. Let's take a look at each of the mental functions affected by growth hormone.

GROWTH HORMONE RESTORES WELL-BEING

Do you love life? Do you wake up in the morning with a delicious sense of anticipation, wondering what the day will bring? Do you have a sense of expanding horizons, a sense that you could do things you've never done before, start a new business, change careers, travel to strange and exotic places, go back to school for an advanced degree? Do you feel that you have the energy, enthusiasm, and passion required to live life to the hilt? If your response to these questions was not an immediate unqualified yes, then think back to when you were a child or a teenager or a young adult. Didn't you *used* to feel this way? One of the things that happens with age is that there is a subtle diminishment in our joy of life. We lose our resilience, our ability to bounce right back after something happens the way that kids do. We become more wary, less willing to try something new whether it is a food or a person or a prospective lover. We feel that it takes too much energy, too much effort. It's so much easier to stay at home at night, zone out in front of the TV, think about cutting back on work or retiring altogether.

SELF-FULFILLING PROPHECY OF AGING

Part of the reason for this feeling is a twist on Descartes, "I think I'm getting old, therefore I am old." Your expectation, your mindset is "I can't do that anymore. I can't be as fast, as smart, as successful." I call this identifying with a model of degenerative disease and death. Picture this scenario: You wake up one day with a pain in your knee and instead of ignoring it or working around it the way you did when you were younger, you worry about arthritis. You look at yourself in the mirror, at your wrinkles, your gray or thinning hair, and worry that maybe that pain won't go away, that it will be with you for the rest of your life. It hurts when you exercise so you stop exercising and then you don't walk as much. You start sitting more, gradually decreasing your activity. When you walk, you do it slowly, carefully, a few steps at a time. You start using a cane, then a walker, finally a wheelchair. You have an expectation of becoming crippled with age and that is exactly what happens.

Growth hormone not only regrows organs and restores bodily

function, it re-invigorates the mind, reversing the attitudes, outlook, and expectations associated with aging. You feel more energetic, sexier, and better-looking, and suddenly there is a whole new gestalt: *"I'm not old anymore."* It is similar to what happens when people who are unhappy with their appearance get plastic surgery. The psychological high that they get is even greater than the lift from the nose bob, ear trim, or tummy tuck.

But beyond the psychological aspect, something else appears to be happening with HGH. It is not just a new outlook on life or an increased sense of well-being, but something that can't be measured on a psychological score sheet. Think of Paulene McNair, being lifted out of her depression and having her appreciation of life restored. R.T. is a forty-eight-year old high-powered lawyer who began taking growth hormone a year ago when he felt run-down and stressed-out. Before starting the program, he felt under constant pressure from his work, the demands of his family, and the difficulty of balancing the two. Today, he says, although his daily life hasn't changed one bit, he has in a big way. "My schedule is more hectic than in any time in my life, yet I feel less stress," he says. "My mood is more elevated than it's ever been. I'm whistling, singing songs while I wash dishes, dancing around. Something is definitely going on. It's almost like being high on something. A natural high." Others tell of starting new careers. Dr. Lee-Benner, who had retired, began lecturing, writing, and practicing medicine again. One of his patients, a famous screenwriter in his sixties, was also ready to call it quits. But on growth hormone, he wrote four screenplays in one year!

To listen to these people describe their experiences and feelings is to be reminded of the movie *Cocoon*. It is like being born into a world where aging no longer exists, where anything is possible, even immortality—in other words, the way they felt as children. The only thing that could account for such a transformation is a change in the neurochemistry, a kind of brain makeover.

THE "LAZARUS EFFECT"

Adult patients who have severe pituitary deficiency and have been deprived of growth hormone for years are evocative of another film—*Awakenings*. Like the victims of viral encephalitis in that movie, many of the GH-deficient adults have been in a kind of suspended animation for decades. "When you look at the patients," says Dr. Bengtsson, "you are struck by the fact that they complain of fatigue, low vitality, low self-esteem, they are socially isolated, their sex life is poor, they are sickly, and are much more apt to be on disability pension than the normal population. They really feel lousy."

On psychological tests, they had more problems than age-matched controls, scoring higher for neurosis and depression. Yet within a few weeks of GH replacement therapy, these patients had increased energy levels, better mood, greater powers of concentration, and improved memory. In some cases, the hormones jolted them awake with such resurrecting force that Bengtsson and his staff termed it the "Lazarus effect." As they continue to stay on the hormone, according to Bengtsson, "They get better and better in the quality of life. The long-term results that we have here are fascinating. The effects are sustained and better and they are happier."

Researchers at St. Thomas Hospital in London led by Peter Sönksen told the same story. Before the addition of growth hormone, the "most striking abnormalities identified were poor energy, emotional lability, low mood and social isolation." More than one-third of the patients "scored within the range of psychiatric disturbance that would normally warrant treatment." These include anxiety, depression, lack of self-control, and impaired emotional reaction. But within one month of starting treatment, many of these patients reported an increase in energy. Six months later, follow-up questionnaires "confirmed the clinical impression of improved psychological well-being, particularly perceived energy and mood."

While there were some physical explanations for this rapid transformation such as increased body water (see below), Bengtsson felt that this could not explain such striking mental and psychological changes. "It was not because you changed your body composition, or your cardiac working capacity," he says. "It had to be a direct effect of growth hormone on the brain and brain function."

They knew that there were receptors in different parts of the brain for growth hormone. But the problem was, how did the growth hormone shots they were giving get to those receptors? The brain is protected from substances floating around in the blood by a tight network of cells known as the blood-brain barrier, and it seemed unlikely that HGH, a large protein composed of 191 amino acids, would be able to pass through.

The only way to tell was to do a spinal tap on some of the patients and look at the cerebrospinal fluid, which runs from the brain through the spinal cord. If there were any growth hormone present in the fluid, it would mean that HGH crossed the blood-brain barrier. To their great surprise, they found that when they gave growth hormone subcutaneously in the leg, there was a ten-fold increase in the cerebrospinal fluid.

They also found the growth hormone injections changed the concentration of certain neurotransmitters (brain chemical messengers). It raised the level of B-endorphin, while it lowered dopamine. B-endorphin has been called the brain's own morphine and is responsible for the "high" feeling that comes with intense exercise. Dopamine can produce feelings of agitation. The same effects, an increase in B-endorphin and decrease in dopamine, is seen in patients who are treated with antidepressants. In other words, it appears that growth hormone replacement has an antidepressive action on the human brain. "Now we have a physiological substrate for the changes we see in the quality of life in these patients," says Bengtsson.

(The relationship between growth hormone and well-being may run very deep. There is a phenomenon known as psychosocial growth failure, in which children who are severely emotionally deprived fail to grow. These children actually have lower growth hormone levels. In other words, love, nurturing, touching, holding—all the things that create a sense of well-being in children—may stimulate normal growth hormone levels, while withholding the ingredients for happiness may literally stunt growth.)

According to the wife of one of Dr. Ullis's patients, growth hormone was more effective in relieving her husband's depression than Prozac was. If growth hormone has a natural antidepressant effect on the brain, it could be a major reason that we are so optimistic and resilient when we are young. With the decline of growth hormone activity with age, we not only lose muscle, bone, thymus tissue, body water, and cell division, we lose our joy of life. But with a program of growth hormone stimulation, we can have it all back again.

GROWTH HORMONE REVIVES ENERGY

One of the first benefits that people on growth hormone experience, often within a few days, is a big boost to their energy. For some it is the complete wake-up call as described by Bengtsson. The effect is most pronounced in people whose energy levels are low to begin with and who are easily tired. In Bengtsson's work with growth-hormone deficient patients, he found that fatigue was a common complaint. "The fatigue reduces their working capacity, influences their professional career, and impairs their leisure activities," he wrote in an article entitled "Long-term Consequences of Growth Hormone Therapy." Researchers at St. Thomas Hospital in London also found that growth hormone had a profound effect in increasing energy and sense of well-being.

In addition to its antidepressant action, growth hormone has an almost immediate effect on body water. The dehydration associated with low growth hormone levels not only dries out the skin but makes people feel sluggish. Like water on a droopy plant, giving growth hormone rehydrates the body tissues and perks up the patient. Second, growth hormone may increase energy and vitality by speeding up cellular metabolism, so that all the workings of the cell are pushed into higher gear typical of the youthful state of the body.

Physicians who have been using growth hormone in their practice find that it is very effective for chronic fatigue syndrome. In one case treated by Dr. Ullis, the patient had high levels of Epstein-Barr virus, which is associated with the syndrome, and poor thymus function. On growth hormone, he has shown remarkable improvement in energy levels and is now able to lift weights and do all kinds of exercise training.

GROWTH HORMONE RELIEVES STRESS

Thierry Hertoghe, M.D., who is an expert on hormonal deficiencies, calls growth hormone, "the hormone of the champion." It is especially good, he says, for removing the anxiety that often accompanies aging. While young people have a greater ability to rebound from the strains of life, older people have lower resistance to stress.

"Over-anxiety or hyper-emotionality is a sign of GH deficiency," he says. "Androgens help both men and women calm down, but you don't get the calmness that you do with GH. Of all the hormones I work with, growth hormone relaxes the individual most."

Hertoghe, who takes growth hormone, finds that it has had the same effect on him. "In the past year, I've been in difficult situations but I haven't experienced the stress that usually goes along with this, like not sleeping well, waking up in the middle of the night worrying, getting up in the morning feeling like you haven't slept. That's all gone away. I know it's not a placebo because when I stop taking the hormone, all these stressful feelings return."

But growth hormone goes beyond having tranquilizer effects. At the same time it relaxes, it improves focus and concentration, and builds self-esteem and self-confidence. "Children who are deficient in growth hormone have low self esteem, are socially isolated and have sharp verbal retorts. But with growth hormone, all these symptoms improved," Hertoghe points out. "You can see the right to heart of a problem and find the solution. I am a leader in many institutions, but I don't get stressed by being a leader. I find new solutions, new creativity. Growth hormone gives an imperial calmness that makes you active and not passive. You act without fear."

GROWTH HORMONE RESTORES DEEP SLEEP

Growth hormone release takes place primarily when we're fast asleep, with its rise and fall following the stages of sleep. A major burst of GH secretion occurs shortly after we fall asleep. It is highest during the deepest slumber of slow wave sleep and lowest during the period of dreaming known as REM (rapid eye movement) sleep. According to Steven Harvey, writing in *Growth Hormone*, "Normal sleep patterns, particularly the maintenance of REM sleep [dreaming], and slow wave sleep may thus be dependent on GH, which is released with the onset of sleep." Sleep deprivation results in lower release of growth hormone, which could explain why we feel so miserable after only one night spent punching the pillow. Adults with severe growth hormone deficiency have abnormal sleep patterns, often spending more time sleeping but less time in slow wave and REM sleep. Giving growth hormone to these adult patients with pituitary deficiency restores normal sleep.

With aging, there is a similar disturbance in sleep patterns. According to University of Chicago researcher Eve Van Cauter, "in the elderly, there is a tremendous reduction in the amount of deep sleep." Older people dream less, awake more frequently during the night, and find it harder to get back to sleep. While the common medical wisdom is that you need less sleep when you're older, many older people complain that they wake up feeling tired. This is not surprising, considering the poor quality of sleep they get.

With growth hormone replacement, patients report great improvement in their sleep and a feeling of being well rested. One woman who has been taking it for a month says she has been able to sleep through the night for the first time in years. Others discover to their delight that they need less sleep. A man in his fifties says that he started waking up after sleeping three or four hours. "At first, I kept trying to get back to sleep. But then I found I was filled with energy and wanted to do things. Now I just get up and enjoy the extra time. And I still feel better all day."

Van Cauter would like to turn things around, investigating whether enhancing slow wave sleep could stimulate growth hormone production. "If restoring deep sleep works," she says, "you reproduce a natural physiological picture of hormone production."

THE VISION THING

Changes in eyesight are one of the most reliable indicators of age. When someone holds a newspaper at arm's length or reaches for half glasses, it's a sure bet that that person is over forty. This is a result of the loss of elasticity in the lens of the eye, starting at age thirty but becoming far worse after the fourth decade, when it causes presbyopia, or farsightedness. The opposite side of the coin, visual acuity or the ability to see those tiny letters at the bottom of eye charts, also falls off, although less sharply, starting around age forty-five. By the age of eighty the distance vision is about a third of what it was in young adulthood. Another problem is that less light reaches the retina, making it difficult to read menus in dimly lighted restaurants or drive at night. By late middle age, many people find themselves stumbling around their bedroom at night, as dark adaptation gets sharply worse. At the same time older people are temporarily blinded when they step into the sunlight, as light adaptation is also failing.

One of the biggest surprises for people on growth hormone is improvement in vision. Some like Howard Turney find that their close vision improves to the point where they no longer need reading glasses. Others report that things come into better focus, as though they were adjusting the fine-tuning on a TV.

Allan Ahlschier, the radiologist, was one of those who found improvements in both near and far vision. Although he still has to wear glasses, he has not had to change his prescriptions since he began using growth hormone three years ago. Some days are better than others, he says, particularly on vacations and weekends when he is not under stress. Two of the most visible benefits, he says, are colors and dark adaptation.

"The colors are magnificent," he says. "I'll be driving at night in Dallas and I'll tell my wife to pull over. I have to look at the colors. It's better than when I was sixteen years old," he says wonderingly. "I didn't have this before growth hormone."

The other big improvement he sees (literally) is in his night vision. "I get up at night, turn the light on and then off, and have instant night adaptation. I can see in the dark. I talked to my ophthalmologist about this and he said, 'slower dark adaptation is a direct sign of aging.'" As an anti-aging consultant, Ahlschier was well aware of that. "The response time to dark adaptation should be listed as one of the biological effects of the aging process along with skin wrinkling, memory capability, and reaction time."

Although the mechanism for increased vision is not known, Ahlschier speculates that it is due to the effects of GH in increasing cell division. Thus the hormone might actually exert a positive metabolic action on the cones within the retina that control color vision and the rods that govern black-and-white vision. "It is probably doing something to both elements," he speculates.

There is some indication that growth hormone can repair damage to the eye. Paulene McNair found that the hole in her vision closed up with treatment (see Chapter 6). And it appeared to save the eyesight of at least one patient. In this case, the problem was macular degeneration in one eye, a major cause of blindness. The man read everything he could on his condition and consulted a number of ophthalmologists, who told him that nothing could be done for him. Finally one eye doctor suggested that he find a physician who would give him growth hormone. When the patient began on the program, his vision in the affected eye was 20/200; after two months of therapy, it was near normal at 20/40.

GROWTH HORMONE AIDS COGNITION AND MEMORY

Two men are talking, one about twenty years of age and the other about sixty.

Young man: Have you noticed a difference between the way you think now and when you were younger?
Older man: I'm much more concerned with the hereafter.
Young man: That's wonderful. You mean you're much more spiritual?
Older man: No, I mean every time I walk into the kitchen, I think what I am here after?

Sound familiar? The failing levels of growth hormone starting in the third and fourth decades of life may be why our minds, like our muscle tone, become less sharp with age. We forget names, lose track of our thoughts, take longer to come up with correct answers, even in the absence of disease. If you want to see just how quickly short-term memory declines, try playing the card game Concentration with a school-age child. (Set out fifty-two cards in rows facedown. The players take turns turning up any two cards. A player who gets a pair goes again. A player who doesn't get a pair returns the cards and tries to remember their location. The winner is the person who matches up the most pairs.) The odds are great that you'll be humiliated.

"Once you hit your 40s or 50s, it's not uncommon to start having little forgetting episodes," says Lee-Benner. "I see a lot of people who write down where they're going on a pad and stick it on the dashboard, so they're remember where they're going." According to noted gerontologist Leonard Hayflick, Ph.D., there is probably a relationship between your age and the number of reminder notes you write yourself.

The decline in growth hormone may be especially important in memory and cognitive performance, according to Jan Berend Deijen and a team of Dutch scientists at the Free University Hospital in Amsterdam. In their study of male patients who were lacking either a number of pituitary hormones including GH, or just GH alone, they found that growth hormone was directly related to impairments in iconic memory (the ability to process a flash of information), short-term memory, long-term memory, and perceptual-motor skills, such as hand-eye coordination. Growth hormone was clearly related to cognitive dysfunction, since the lower the levels of IGF–1 (an indication of growth hormone production), the lower the patient's IQ and education level. Most of these patients, the researchers point out, had been deficient in growth hormone from birth. This could mean that the lack of GH in childhood prevented the full development of their brains. In such cases, they believe, GH replacement may not help their cognitive problems. But if the cognitive problems are caused by abnormal metabolism of brain cells as a result of GH deficiency, then growth hormone replacement could improve the metabolism and increase their memory, concentration, and thinking power, they say.

"We are just beginning to scratch the surface of the importance of growth hormone in cognition, memory and brain function," says Bengtsson. His own research and that of others show that growth hormone replacement in GH-deficient adults improves cognition (our ability to think and reason) as well as memory. In children it has been shown to have remarkable effects. Growth hormone–deficient children who were started on GH therapy before age five have increased head size, have higher IQs, and did better at school.

Patients on growth hormone recount being more alert, less forgetful, and better able to concentrate. Some of them are rediscovering the nimbleness of youth with greatly speeded up reaction times. R.T., the forty-eight-year-old lawyer, recently tested his reflexes against a woman in her early twenties and two men who were twenty-one and twenty-two. The test involved a stardard measurement of aging—catching a falling ruler (see Chapter 13). "We did it a number of times," he says, "and I was twice as fast as these young people." The kick to the brain has probably also made it possible for Bedford King to learn an entirely new field of commodities trading at the age of seventy-four (see Chapter 6).

BIGGER, BETTER BRAINS?

There is no doubt that growth hormone plays a major role in the brains of newborns and young animals, stimulating the growth of neurons, the message-carrying cells of the nervous system, and the glial cells, which support the neurons, and expanding the brain and skull size. In rats it has been found to increase the cells of the myelin sheath that insulates the central nervous system. (It is the destruction of the myelin sheath that causes the symptoms of multiple sclerosis.) And it spurs the synthesis of RNA in brain cells grown in the laboratory. But the question is, can GH affect the fully formed brain of an adult human? The answer appears to be yes, that here too GH can stimulate cell division, repair, and rejuvenation.

REVERSING BRAIN SHRINKAGE

In fact, growth hormone may actually be regrowing the brain as it regrows the other organs of the body. The truth is that your brain is actually shriveling up with age. Your brain reaches its largest size at age twenty, when it weighs about three pounds. Thereafter it starts to shrink, until by age ninety, it weighs about 10 percent less than it did at its maximum. There is some dispute about *why* it is getting smaller. Some gerontologists argue that you are losing brain cells at a rate of about 50,000 to 100,000 a day. But according to neuroscientist Robert Terry at the University of California in San Diego, the cells aren't dying, just shrinking. What is being lost, he says, is the dendritic connections between cells, the branchlike projections of nerve cells that spread out in every direction to hook up with other nerve cells. The brain cannot sprout new neurons the way the skin, blood, or intestines can grow new cells. But when stimulated with growth factors, neurons can regrow dendritic connections. And other brain cells, like glial cells, which nourish the neurons, can also spring anew.

In a classic series of experiments, Marian Diamond, Ph.D., professor of anatomy at the University of California at Berkeley, showed that putting aged rats into an enriched environment—outfitting their cages with ladders, wheels, bells, and other rodent goodies—caused new dendrites to sprout. In effect, she changed the brains of old rats into young ones, increasing the growth of their cortex by 5 percent.

When Diamond carried out her experiments, she upended the prevailing belief that once the brain reached adult size, no further growth could take place. But the same growth factors that heal and repair wounds, and grow new skin, collagen, muscle, blood, and other tissue, also grow peripheral nerves and dendrites. Elderly rats who had nerve growth factor injected into their brains were able to learn to swim through a maze in water, while untreated rats were not able to do so. It may be that growth hormone is stimulating growth factors in the brain, promoting the sprouting of new dendritic connections and regenerating brain cells.

PREVENTING BRAIN AGING AND DISEASE

One of the spectacularly exciting uses of growth hormone or IGF–1 may be to prevent and treat the effects of brain aging. In an experiment that has momentous implications for brain injury, stroke, aging, and neurodegenerative disease, a team of scientists in New Zealand showed that IGF–1 can stop the death of cells in the brain. Barbara Johnston, Peter Gluckman, and their colleagues at the University of Auckland found that injections of IGF–1 given *two hours after* brain injury in fetal lambs rescued the damaged neurons. And salvaged cells that would otherwise have died during apoptosis, the programmed cell death that is believed to cause the loss of brain cells for up to three days after the original injury. The treatment was effective in stopping cell death throughout the brain, including the cortex and hippocampus, the areas associated with thinking and memory, respectively, and in the striatum, the part which plays a role in Parkinson's disease in humans. IGF–1 replacement also reduced the rate of seizures in the brain-damaged animals.

There are other recently developed therapies for brain damage, such as free-radical scavengers and calcium channel blockers that can prevent neuronal death but only when they are given *before* the damage occurs. This does not make them practical for real-life situations. Many of these agents also have toxic side effects. IGF–1 is the only treatment that was effective when given after the injury and had no adverse effects. At this point it has to be injected directly into the brain, but unpublished experiments show that it might be possible to give it as a spinal injection, which would reach the brain.

The New Zealand researchers suggest that IGF–1 might be given to counter the effects of hypoxia (lack of oxygen) during birth, which can leave a baby with permanent brain damage. But if IGF–1 or growth hormone can stop apoptosis, the programmed death of cells, it opens up a world of undreamed-of possibilities. The programmed death of cells after a heart attack leaves a corpse of tissue that will never beat again. The programmed death of cells in the brain after a stroke can destroy the ability to walk, or to use an arm, or to talk, or to think clearly ever again. The programmed death of cells in aging can make it harder to concentrate, react quickly, recall names, dates, facts, what you were doing just the other day. The programmed death of cells in the corpus striatum of the brain can cause Parkinson's disease. It may also play a role in other neurodegenerative diseases, including multiple sclerosis, muscular dystrophy, Lou Gehrig's disease, and Alzheimer's disease. With IGF–1 or growth hormone, we may for the first time have a weapon against death itself, at least on the cellular level.

By maintaining youthful levels of growth hormone as we age, we may stem the loss of neurons and keep the brain youthful and functional throughout our life span. Cass Terry, professor of neurology at the Medical College of Wisconsin, believes that growth hormone may have the same beneficial effect on the brains of older people that it does on the heart. "As you get older, your memory starts to decline. We see those patients all the time in our clinical practice and they are not demented. But they are intelligent enough to know that they are not as good as they used to be."

Terry plans to study if growth hormone could slow or reverse this decline with age. "If the memory loss that occurs with age is due to the drop off of neurons, then theoretically you may be able to slow that process with growth hormone," says Terry. It might also be helpful for neurological conditions due to aging." As one example, he points to Parkinson's disease, which is caused by loss of dopamine-containing cells in the brain. "People who develop Parkinson's," he says, "probably have a lower reserve account of dopamine-containing cells, so that some people get it earlier than others. Maybe growth hormone could slow that process as well."

CAN HGH TREAT NEURODEGENERATIVE DISEASE?

Sam Baxas, M.D., of Baxamed Clinic in Switzerland has been treating patients with Parkinson's, Lou Gehrig's disease, Alzheimer's, and other degenerative muscular and nervous diseases for twenty-two years with cell therapy injections—extracts of cells and growth factors from unborn calves and sheep. Seven years ago, Baxas added growth hormone replacement to his regimen for those whose levels of GH were deficient. "The combination has been really explosive. We are healing just about everything—heart disease, chronic fatigue syndrome, multiple sclerosis, muscle wasting diseases, Lou Gehrig's disease (amyotrophic lateral sclerosis), Parkinson's, lupus erythematosus, all the autoimmune diseases you get when your immune system goes haywire. We are very successful in treating AIDS patients. We are healing them."

Baxas believes that growth hormone and cell therapy work by rejuvenating the cells of the body and brain. He calls them "antithanatopsis," for the Greek word for death. "Growth hormone and cell therapy stop aging in the cell," he says. "If you take a petri dish with cells in it that have stopped dividing and put cell therapy or growth hormone in there, the cells start dividing again. With growth hormone and cell therapy, we can slow down and even stop the aging process, even turn it back a little."

To date, there have been no double-blind studies on the use of growth hormone or IGF–1 in the treatment of neurodegenerative disease. But there are intriguing signs that Baxas and others who claim positive results may be on the right track. IGF–1 has been shown to regenerate nerve tissue—whether damaged by injury or illness. In studies of nerve cells in culture and in animals, it has repaired peripheral nerves, the ones that innervate the arms and legs, as well as the nerves of the spinal cord and central nervous system. This makes it a likely candidate for degenerative disorders like Lou Gehrig's disease. Nerve growth factors, which are stimulated by GH, have stopped the progression of this disease in mice. (For more on IGF–1 and nerve growth factors, see Chapter 21.) Growth hormone is found in brain cells that control motor activity, and GH therapy normalizes the impaired motor activity of dwarfed mice, suggesting that it may be of value in treating Parkinson's patients.

Luis Carlos Aguillar and Jose Cantu of the Institute for Research of Cell Regeneration in Guadalajara reported on three patients with Parkinson's treated with intramuscular injections of nerve growth factor and other growth factors. A forty-year-old became free of the uncontrollable movements, a sixty-year-old was much improved, and a seventy-six-year-old was somewhat improved. Growth hormone also helps motor activity by stimulating the growth of myelin sheath on nerve cells, making it a potential treatment for multiple sclerosis. In Alzheimer's disease, there is a loss in a number of neurotransmitters, particularly acetylcholine and noradrenaline, which stimulate growth hormone. A number of investigators have suggested that GH may be useful for the treatment of this disease. Hertoghe believes multihormone replacement, including growth hormone, can help Alzheimer's disease. When we are ninety, our brains have shrunk down to the size they were when we were three years old, he says. He believes that growth hormone restores the more natural form and metabolism of the brain by adding water to the tissues and removing fat. With Alzheimer's you also get a shrinking of the brain, he says, which can be shown on magnetic resonance imaging (MRI). He is now treating several Alzheimer's patients with hormone replacement therapy including growth hormone and is planning to follow their progress with MRI.

TREATING HOPELESS PATIENTS

A seventy-one-year-old woman in the later stages of Alzheimer's disease is tearful, withdrawn, confused, unable to remember where she is going, the day of the week, or her sister's name. Six months after treatment, she is cheerful, related, able to take of herself, and able to relearn old tasks as well as learn new ones.

Patients with Parkinson's who move in slow shuffling steps, unable to pick up their feet or turn around easily, are running down the hall and jumping into the air after one injection. Gordon Cooper, the famous astronaut, now suffering from Parkinson's, is captured on videotape, running, kicking up his heels, and bending from the waist and touching the floor, knees slightly bent, as he demonstrates his freedom from his symptoms, also after one treatment.

These are the patients of Chaovanee Aroonsakul, M.D., of the Alzheimer's-Parkinson's Disease, Diagnostic and Treatment Center in Napierville, Illinois, a suburb of Chicago. Dr. Aroonsakul holds seven U.S., Canadian, and Australian patents for the diagnosis and treatment of neurodegenerative diseases, such as Alzheimer's, Parkinson's, and senile dementia, and aging-related degeneration in adults with growth hormone deficiency symptoms. Her treatment method involves hormonal replacement using such things as estrogen, testosterone, and chorionic gonadotropin, a hormone produced by the placenta in pregnancy, which has identical activity as both the male and female sex hormones. In the later stages of disease, she uses human growth hormone.

Since she has neither conducted clinical trials nor published her findings, it is impossible for us to evaluate the validity of her treatments. Her videotapes feature what sound like unscripted interviews with the patients and family members. In the case of the woman with Alzheimer's, according to the videotape, her evaluation included a rating scale developed specifically to assess psychosocial behavior in the elderly—the Sandoz Clinical Assessment Geriatric Orthogonal Factors (SCAG). The higher the score, the worse off the patient. In February 1985, at the time of her referral to Dr. Aroonsakul, she scored 12 in agitation-irritability; 21 in mood depression, 27 in cognitive dysfunction, and 15 in withdrawal, for a total of 75, indicating advanced disability. Six months after treatment her scores were 7, 11, 11, and 7, respectively, for a total of 36. She was still confused but smiling, and somewhat better. By January 1986 her scores were 0, 3, 5 and 0, for a total of 8, an incredible improvement for a disease in which the inevitable course is a downhill slide into a vegetative state and death.

Dr. Aroonsakul claims that she has many such cases. In her patent applications, she says that her treatment methods have reversed the degenerative nature of the diseases and have restored patients suffering from the diseases to more normal and productive lives, with the alleviation of many of the symptoms of the diseases. With continued treatment, she says, "There have not been found any diminution of efficacy of the treatment, nor any serious contraindications and adverse side effects."

What hormone replacement does, according to Aroonsakul, is reverse the processes of cellular atrophy and deterioration that leads to these diseases. Different parts of the brain are affected in different diseases. For instance, in Parkinson's it is the area of the basal ganglia, where the dopamine-containing area are located, while in Alzheimer's it is primarily the cortex, which contains the cells vital to memory. While cellular atrophy occurs with aging in normal brains, it is speeded up in these diseases, for reasons that are not yet understood. But, she points out, the very fact that cells are atrophying means that the enzymes that make the protein machinery of the cells—the DNA, RNA, and enzymes—have broken down. Replacement of growth hormones and the sex hormones, which also stimulate growth hormone, restarts the protein machinery and revitalizes the nerve cells and the synaptic connections between them, she says. Another theory of how her treatment works is based on the idea that the brain cell metabolism slows with age. This allows amyloid deposits to accumulate in the brain cells. Amyloid is an abnormal protein that is found in large quantities in the brains of Alzheimer's victims. By speeding up brain cell metabolism, the theory goes, the accumulation of amyloid is prevented.

CAN ESTROGEN REPLACEMENT STOP ALZHEIMER'S?

Two long-term studies at the University of Southern California shows that Aroonsakul might really be on to something. In the first study, Victor Henderson and his associates found only 7 percent of a group of 143 women with Alzheimer's disease were on estrogen replacement, while 18 percent of a nondemented control group of 92 matched for age and education used estrogen postmenopausally. The second study, which looked at almost 9,000 women living in Leisure World, a retirement community in Southern California, found a 30–40 percent reduction in risk of developing Alzheimer's for women on estrogen replacement compared with those not receiving hormone replacement. In an interview with the *Chicago Tribune*, Henderson said, "I don't want to oversell these results. We still consider them preliminary. But they suggest that estrogen replacement might be useful for preventing or delaying the onset of this dementia in older women." And it may be beneficial in treating women who already have the disease. Henderson and his colleagues found that women who had Alzheimer's and were receiving estrogen did better on a test for mental function than the women who had Alzheimer's and were not on estrogen. In addition, two small clinical trials have found estrogen to be beneficial in treating Alzheimer's.

It may be that a program of hormone replacement with growth hormone, GH agonists, and other hormones that decline with age will not only protect the brain and body as we age, but also prevent, delay, or reverse the most dreaded aspects of old age—the neurode-generative diseases that delete our memories, our personalities, our very identity as human beings.

So far in this book, we have provided you with the scientific information that demonstrates why growth hormone is the hor-mone of youth. In the next section, we will give you all the tools you need to design your own personal program of growth hormone enhancement.

PART 2

The Growth

Hormone

Enhancement

Program

Determining Your Need: A Self-Test

The first step in designing a growth enhancement program is determining whether you are actually deficient in GH for your age. There are a number of simple tests you can perform on yourself that may indicate a deficiency of GH and can also tell you how well you are aging. The only ways to know for certain if you are deficient in growth hormone is to have a laboratory test your levels of IGF–1 or do a direct determination of your GH levels as described in the next chapter.

DIAGNOSING GH DEFICIENCY

Low growth hormone levels, as we have said a number of times in this book, mimic the effects of aging, so that just a general sense that you are going downhill may be sign of deficiency. In fact, this is the most common complaint of patients who go on GH replace-. ment. They talk about a lack of energy, they say that life seems less exciting, they feel a decline in libido and sexual performance. The same symptoms are seen in adult patients who develop pituitary disease.

The psychological and emotional symptoms associated with low growth hormone levels include a reduced sense of well-being; low energy, vitality, and capacity for work; emotional lability, including mood swings, anxiety, and depression; and increased social isolation. Important physical signs are increased body fat, especially around the waist (apple-shape rather than pear-shape), decrease in muscle mass, thin, wrinkled, or prematurely aged skin. The following signs and symptoms of GH deficiency are listed in *An Introduction to Growth Hormone Deficiency in Adults* by Bengtsson.

SYMPTOMS

Impaired psychological well-being and quality of life with:
    poor general health
    impaired self-control
    lack of positive well-being
    depressed mood
    increased anxiety
    reduced vitality
    reduced energy
    impaired emotional reaction
    increased social isolation

SIGNS

    reduced lean body mass
    reduced extracellular fluid volume
    reduced bone mineral density
    increased body fat
    increased waist:hip ratio
    decreased HDL cholesterol
    increased LDL cholesterol
    reduced renal plasma flow
    reduced basal metabolic rate
    reduced muscle bulk
    reduced muscle strength
    reduced exercise performance
    reduced anaerobic threshold

The following self-test is designed to help you determine whether your hormonal health is declining and you should seek to elevate your levels of GH and other hormones (see Chapter 15) either with the Growth Hormone Enhancement Program (Chapters 16 through 18) or with HGH injections (Chapter 19). This is a basic screening test only. A decision to start a hormone replacement program or HGH therapy should only be done after consulting with your physician (see Chapter 14 on choosing a doctor) and having your hormonal levels tested. Take the self-test again in three months if you have gone on growth hormone replacement or stimulation with growth hormone releasers to tell how well the program is working for you.

ARE YOU A CANDIDATE FOR HORMONAL REPLACEMENT THERAPY?

Today, as compared to 10 years ago:	*If yes*
1. Do you often feel tired?	+1
2. Do you feel happy most of the time?	−2
3. Do you often go through mood swings?	+2
4. Do you anger easily?	+2
5. Are you depressed often?	+1
6. Do you often feel anxious or stressed out?	+1
7. Do you feel you work too hard?	+2
8. Do you look forward to retirement (not to pursue an activity but to do less)?	+2
9. Do you keep in touch with friends?	−1
10. Do you maintain an interest in sex?	−1
11. Is your sex life declining?	+2
12. Do you have trouble falling or staying asleep?	+2
13. Do you feel well rested after sleep?	−1
14. Do you find yourself forgetting things?	+2
15. Do you find it harder to think clearly?	+2
16. Do you use memory aids (e.g., lists)?	+2
17. Do you have problems concentrating?	+2
18. Are you in poor physical shape?	+2
19. Are you more than 20 percent above your ideal weight?	+2
20. Is it very difficult for you to lose weight?	+1
21. Have you developed a spare tire or love handles?	+1
22. Does your musculature look youthful?	−2
23. Do you feel your overall health is good?	−2
24. Do you often get colds or feel sick?	+2
25. Do you commonly feel aches or pains?	+1
26. Is your blood cholesterol over 200?	+1
27. Is your blood cholesterol over 240?	+2

28. Men—is your HDL less than 45?
    Women—is your HDL less than 55?   +2
29. Is your blood pressure normal?   −2
30. Has your vision noticeably deteriorated?   +1
31. Do you have frequent urination?   +1
32. Do you have digestive problems?   +1
33. Does the skin on your face, neck, upper arms,
    and abdomen appear to hang?   +2
34. Do you think you look older than your age mates?   +1
35. Do you have cellulite on your thighs?   +1
36. Do you need haircuts less frequently?   +1
37. Does it seem to take a long time for cuts and
    bruises to heal or for wounds to close?   +1
38. Is it getting harder to exercise?   +2
39. Do you seem to have less strength for gripping
    or lifting?   +2
40. Is your endurance less?   +2
41. Is your breathing more labored when you exercise
    hard?   +3
42. Do you find the longer you live, the better you
    feel about life?   −2

Ages 45 to 54   +1
Ages 55 to 64   +2
Ages 65 and above   +3

Total   _____

14 and below: You are doing well and your complaints are well within normal range of daily living.

15–22: The Growth Hormone Enhancement Program may help forestall some of the problems of aging.

23–30: Hormonal replacement therapy with HGH or the Growth Hormone Enhancement Program may reverse the problems of aging you are already encountering. Schedule a visit to your doctor and have your IGF–1 levels checked.

31 and above: Run, do not walk, to an anti-aging physician (but not so fast that you fall and break a hip or get a heart attack). Chances are your levels of growth hormone are severely deficient. Hormonal replacement therapy including HGH may be of great benefit.

Even if you feel satisfied with your general physical and mental health, you may wish to check the self-test biomarkers below. These are measurements of your biological age versus your chronological age, that is, whether your body is younger or older than your age at your birthday would indicate. No matter how lucky we are in our choice of ancestors, how healthful our lifestyle practices, how pure

and wholesome our thoughts, we still travel an "inexorable" downhill path, some of us more quickly than others. Stimulation of the
GH and IGF–1 system can to some degree reverse this course.

DIAGNOSING YOUR AGING RATE

Here are four tests adapted from *The 120-Year Diet: How to Double
Your Vital Years* by Roy Walford, M.D. Following is a chart that shows
the average values of these tests for three different age groups.

1: *Skin elasticity test.* Loss of skin elasticity starts to be significant around age forty-five and is a result of the underlying deterioration of the connective tissue, such as collagen and elastin, under
the skin surface. This loss of skin tone and turgor contributes to the
wrinkling and loose skin around the jowls and neck. To do the test,
pinch the skin on the back of your hand between the thumb and
forefinger for five seconds. Then time how long it takes to flatten
out completely. Average rates per age: between forty-five and fifty
years—five seconds; sixty years, ten to fifteen seconds; seventy
years—thirty-five to fifty-five seconds. There are large individual
variations, so don't take this test too seriously. Use it mostly as a
gauge to see how you are doing. (See Chein's claim of his own progressive skin tightening with GH, Chapter 10.)

2: *Falling ruler test.* This is a test of your reaction time, which
falls off sharply with age. Slow reaction time is what kills off old
animals in the jungle and old people in city streets when they fail to
step out of the way of a quickly turning car. To do the test, buy an
eighteen-inch wooden ruler. Have someone suspend the ruler by
holding it at the top (larger numbers down) between your fingers.
The thumb and middle finger of your right hand (use your left
hand if you're a lefty), should be three and a half inches apart,
equidistant from the eighteen-inch mark on the ruler. The person
lets the ruler go without warning and you must catch it between
your fingers as quickly as possible. Do this three times and average
your score. (For instance, if you catch it at the three-inch, six-inch,
and six-inch marks, your score is three plus six plus six, divided by
three, which equals five.) The average score at age twenty is the
eleven-inch mark, descending to six-inch mark at age sixty. (Growth

hormone may reverse reaction time in this test as shown by the experience of R.T. in Chapter 12.)

3:   *Static balance test.* This is a test of how long you can stand on one leg with your eyes closed before falling over. According to Walford, this is the best of the "do-it-yourself biomarker measurements." There is a 100 percent decline on average from age twenty to age eighty. To do the test, stand on a hard surface (not a rug) with both feet together. You should be barefoot or wearing an ordinary low-heeled shoe. Have a friend close by to catch you if you fall over. Close your eyes and lift your foot (left foot if you are right-handed and right foot if you are left-handed) about six inches off the ground, bending your knee at about a forty-five-degree angle. Stand on your other foot without moving or jiggling it. Have someone time how long you can do this without either opening your eyes or moving your foot to avoid falling over. Do the test three times and take an average. A young person can usually hold a one-legged, eyes-closed stance for thirty seconds or more, while an old person usually falls over within a few seconds. Check the biomarker chart that follows for values at different ages.

4:   *Visual accommodation test.* This test shows you why most people by the time they are forty-five are reaching for their half glasses or bifocals. With age, the lens of the eye becomes progressively less elastic, resulting in presbyopia, or nearsightedness. While this is not as accurate a test of visual accommodation as your eye doctor can do, it will give you some idea of the effect of age on your vision. To do the test, slowly bring a newspaper to your eyes until the regular-size letters start to blur. Have someone measure the distance between the eyes and the paper with a ruler. At age twenty-one, this distance will be within four inches; at thirty years, within five and a half inches; at age forty, nine inches; at age fifty, fifteen inches; at age sixty, thirty-nine inches. Do this either without glasses or with glasses corrected for distance (not reading glasses). It is not how well you can read the print, but at what point it starts to blur. The graph below illustrates how sharply visual accommodation falls off after age forty.

Near Vision

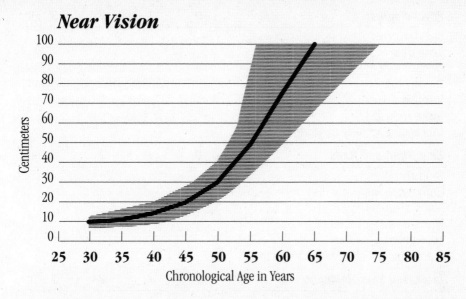

The shaded area represents the distance at which 90% of the population can see near objects clearly.

BIOMARKERS OF AGING TESTS FOR THREE AGE GROUPS				
Test	**20–30**	**40–50**	**60–70**	**Units**
Skin elasticity	0–1	2–5	10–55	seconds
Falling ruler	12	8	5	inches
Static balance	28	18	4	seconds
Visual accommodation	4.5	12	39	inches

KEEP A RECORD OF YOUR PROGRESS

This worksheet is adapted from the High Progress Questionnaire of the Renaissance Rejuvenation Centre in Cancun, Mexico, with a few additions of our own. (Enter 0–3, where 0 is no change, 1 is little change, 2 is a moderate change, 3 is a large change.) Test yourself after two months and again after six months.

Psychological	0	1	2	3
Energy levels				
Increased mental alertness				
Increased sense of well-being				

Psychological	*0*	*1*	*2*	*3*
Less anxious				
Less depressed				
Deeper sleep				
Dream more				
Improved recall				
More focused				
Increased activity at work				

Skin, Hair, and Nails	*0*	*1*	*2*	*3*
Skin thicker				
Skin more supple—less dry				
Nails stronger				
Nails growing faster				
Hair growing faster				
New hair in previously bald areas				

Increased Skin Tightening	*0*	*1*	*2*	*3*
Under upper arms				
Abdomen				
Thighs				
Less sag under chin				
Skin on neck				
Facial skin				

Immune System	*0*	*1*	*2*	*3*
Fewer allergic reactions				
Faster healing (gums, cuts, surgery, etc.)				
Fewer colds				

Digestive System	*0*	*1*	*2*	*3*
Less indigestion				
More regular bowel movements				
Bulkier stools				

Urinary System	*0*	*1*	*2*	*3*
Less frequency				
Less getting up at night to urinate				
Less urgency				

Muscles	*0*	*1*	*2*	*3*
Stronger abs				
Thigh muscles firmer				
More youthful body contour				
Love handles and belly disappearing				

Extremities	0	1	2	3
Hands appear younger				
Brown spots fading				
Veins less prominent				
Feet appear younger				

Sex and Reproduction	0	1	2	3
Increased sex drive				
Firmer erections (men)				
Greater frequency of sexual relations				
Increased enjoyment				
Fewer problems with menopause (women)				

MEASURING BODY COMPOSITION AND STRENGTH

If your personal Growth Hormone Enhancement Program is working, you should see significant changes in your body composition and strength levels. You may even grow taller! The visible, measurable alterations in your physique will give you the motivation you need to continue the diet and exercise part of the program that will ensure the greatest success.

On a separate worksheet, record the measurements listed below at baseline (when you begin the program) and at three months, six months, one year, and eighteen months. If you are on growth hormone replacement, you can expect to lose on average about 8–10 percent of your body fat and gain 4–8 percent in muscle mass in about six months. Keep a record of your weight, but remember that because muscle weighs more than fat and your muscle mass is increasing while your fat mass is decreasing, your overall weight may not change or may even go up! Since you are recording changes after three months and then at six-month intervals, you will cancel out most of the variability you would have if you did this on a daily basis. Still, it is best to have an accurate scale or use a doctor's scale and to weigh yourself at the same time of the day and, if possible, nude. A good time is when you wake up or before your largest meal.

To check your height, do it the way families have long measured the growth of their children. Stand with your back to a wall, stretching yourself as tall as you can while you try to flatten yourself against it. Have someone mark the wall where the top of your head meets it. Check yourself in three months at the same place, putting in the new mark. Since this is a measure of bone density in the spine as well as retention of body water, don't expect any change for a year to a year and a half, since it takes that long for new bone growth to accumulate.

For changes in body contour, continue measuring your waist and hips and getting the ratio, as described below. You can gain a fairly good idea of your fat-to-lean ratio by using skin-fold calipers. Calipers are simple to use once you get the hang of it. You make two skin-fold measurements as shown in the drawings below for women and men. Women use the skin fold on the top of the hipbone and on the back of the arm, and men measure the skin fold on the front of the

Skin Fold Measurements

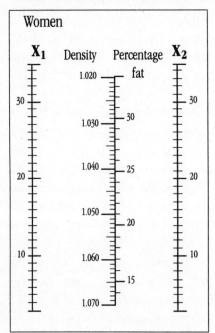

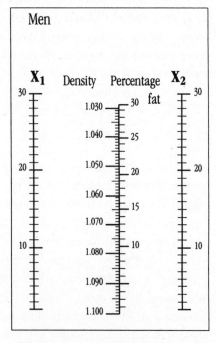

Young women. X_1 = skin-fold thickness (mm) and X_2 = back of arm skin-fold thickness (mm). Adapted from Sloan and de V. Weir, "Nomograms for Prediction of Body Density and Total Body Fat from Skin-fold Measurements," *J. Appl. Physiol.* 28(2):221–22 (1970).

Young men. X_1 = skin-fold thickness (mm) and X_2 = back of arm skin-fold thickness (mm). Adapted from Sloan and de V. Weir, "Nomograms for Prediction of Body Density and Total Body Fat from Skin-fold Measurements," *J. Appl. Physiol.* 28(2):221–22 (1970).

thigh and below the shoulder blade. Do each measurement three times. The three measurements should not differ by more than 1 percent. If they vary more than that, you'll have to be more careful in your measurements. Try to measure yourself in the same place each time. If you wish to record it on a weekly basis, which we recommend, mark the two spots where the calipers go with an indelible laundry pen, which won't wear off until new skin grows and covers them. (This could be a good way to see if your skin is growing any faster as well.) Next plot the two skin-fold measurements on the graphs shown below for men and women. Draw a line between the two measurements and where they crossed the middle and, voilà, you have your percentage of body fat. (See example given in the figure below.) By multiplying your total body weight by your percentage of body fat, you get the number of pounds of fat on your body.

Example of Skin-Fold Measurements

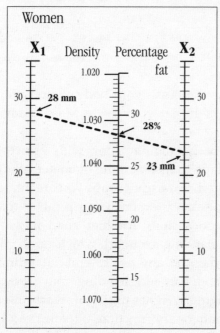

In this example, X_1 = skin-fold thickness is 28 mm and X_2 = back of arm skin-fold thickness is 23 mm. Drawing a line between the two measurements on the charts gives the average percentage body fat of 28%.

Are You at Risk?

A BMI of 27 or over means your weight increases your chance of developing health problems like diabetes, heart disease and certain cancers. Find your height on the chart, then look across that row and find the weight nearest your own. The number at the top of the column is your BMI.

BMI	19	20	21	22	23	24	25	26	27	28	29	30	35	40
	Weight in pounds								←— Risk zone —→					
5'	97	102	107	112	118	123	128	133	138	143	148	153	179	204
5' 1"	100	106	111	116	122	127	132	137	143	148	153	158	185	211
5' 2"	104	109	115	120	126	131	136	142	147	153	158	164	191	218
5' 3"	107	113	118	124	130	135	141	146	152	158	163	169	197	225
5' 4"	110	116	122	128	134	140	145	151	157	163	169	174	204	232
5' 5"	114	120	126	132	138	144	150	156	162	168	174	180	210	240
5' 6"	118	124	130	136	142	148	155	161	167	173	179	186	216	247
5" 7"	121	127	134	140	146	153	159	166	172	178	185	191	223	255
5' 8"	125	131	138	144	151	158	164	171	177	184	190	197	230	262
5' 9"	128	135	142	149	155	162	169	176	182	189	196	203	236	270
5' 10"	132	139	146	153	160	167	174	181	188	195	202	207	243	278
5' 11"	136	143	150	157	165	172	179	186	193	200	208	215	250	286
6'	140	147	154	162	169	177	184	191	199	206	213	221	258	294

(Height — left margin label)

You can also follow changes in your body composition by using your weight and height to get your Body Mass Index (BMI), as shown in the graph.

If you are doing bodybuilding along with growth hormone stimulation, use your tape to measure the diameter of your biceps, calf, chest, neck, or any other muscle groups you are interested in.

For grip strength, use an inexpensive grip meter that is available at any sporting goods store. Another good measurement of muscle strength is the standing arm curl, which can be done using the mechanized equipment at any health club. Standing up, with your arms straight, grab a bar, and, bending your elbow, curl it to your chest. Keep changing weights until you reach the maximum you can curl. To test pulmonary function, all you need is a candle and a book of matches. Light the candle and try blowing it out with your mouth open. Keep moving the candle farther away until you can't blow it out any longer. Record the distance. This is a very rough measurement of your vital capacity, which many gerontologists consider the strongest predictor of life expectancy.

	3 months	6 months	1 year	18 months
Weight				
Height				
Skin-fold thickness				
X_2				
X_1				
Percent body fat				
BMI				
Grip strength				
Standing arm curl				
Lung function				
Muscle groups				

JUST SAY *NO* TO AGING

The problem with aging is that you can adapt to less-than-optimal health without realizing it. It's like driving around in an old car and compensating for its shaky brakes or steering wheel by pressing down harder on the brake pedal or turning the wheel more than you would normally need to. It's not until you've had the car tuned up and overhauled that you notice the difference. And that's the way many people feel when they have experienced the newly over-hauled body made possible through a total program of growth hormone stimulation.

If you wish to keep aging on schedule, then we suggest that you stop reading at this point. But if want to opt for optimal health and feeling better than you ever have in your life, turn the page.

1 4

Finding an Anti-Aging Doctor

WHY YOU NEED ONE

There are several compelling reasons why you should not undertake to treat yourself without medical supervision. First, only a doctor can prescribe growth hormone or growth hormone–releasing prescription drugs as well as some of the hormone replacements covered in the next chapter. Second, you will need a doctor to monitor how well the Growth Hormone Enhancement Program is working by measuring the levels of IGF–1 as well as any other hormones you wish to replace. Third, you may wish to follow the effect that growth hormone is having on your various risk factors for disease, such as heart disease, diabetes, and osteoporosis. These require sophisticated measurements that can only be done by a doctor and a reputable laboratory. Fourth, there can be adverse or uncomfortable side effects with any treatment, including the nontoxic GH-releasers listed in Chapter 16. Finally, we are all of us different from one another with a unique set of genes, biochemistry, risk factors, and medical and familial history. Only a physician—and preferably a physician specializing in anti-aging medicine—in consultation with you, can determine which program works best for you.

HOW TO CHOOSE AN ANTI-AGING SPECIALIST

To become a physician, you have to take the Hippocratic Oath, in which you swear to "do no harm." But there are many ways to interpret that phrase. As all Catholics are aware, there are sins of omission and sins of commission. Cutting off the wrong leg of a diabetic or operating on the wrong side of the head in a patient with a brain tumor as happened recently in two New York hospitals is clearly doing harm. But what about not keeping abreast of the latest diagnostic treatment or advances? Or failing to inform a patient of lifestyle changes or options that could drastically lower risk of disease? Or taking a authoritarian, I-know-better-than-you attitude that effectively cuts off all questions about alternative measures and treatments?

All these "sins of omission" may end up doing you harm in terms of accelerated aging, disease, and death.

If you care about achieving optimal health and the longest possible functional life span, you want a specialist in anti-aging medicine or, at the very least, someone who has an orientation in preventive medicine and maintaining the highest-quality health.

Start by looking at who your primary health caregiver is. Ask him or her the following questions:

1: *Do you think we can delay aging or the diseases associated with aging?* While there is some dispute about what aging actually is (see Chapter 5), a physician who practices anti-aging medicine believes that we can postpone or reverse many of the biochemical changes associated with later life disease, such as heart disease, cancer, stroke, and diabetes. Beware if your physician says something like: "Well, you're just getting old, what can you expect? People at your age get aches and pains, start slowing down, have trouble getting it up, etc. I know a lot of people who are sixty and can't get out of the bed in the morning." My advice: Get out of that person's office while you're still able to move.

2: *What do you do for your own health?* Physicians who have an anti-aging orientation practice what they preach—exercising regularly, taking high-quality multivitamin and mineral formulas along with supplements of antioxidants and other compounds that reflect their own research and that of others. It is interesting to note that every doctor who prescribes growth hormone replacement uses either HGH or GH-releasers on himself or herself if the IGF–1 levels indicate a deficiency. Look for a doctor who spends money and time on a personal anti-aging program and is proud of the fact.

3: *How many patients are you actively treating in anti-aging regimens?* You don't want someone who sees anti-aging medicine as a passing fad but someone who is actively practicing it. While a doctor's practice is made up mostly of people who come in when they're in extreme discomfort, a physician who does anti-aging medicine should have at least thirty to forty patients who are committed to a program of optimal health.

4: *How do you keep up with latest advances?* Medical knowledge increases exponentially, doubling every 3.5 years. Nowhere is this more true than in the high-technology arena of anti-aging medicine, where there have been tens of thousands of scientific articles on growth hormone, melatonin, and DHEA just in the past few years. In addition to mainstream scientific journals, doctors should subscribe to periodicals that cull the latest anti-aging advances, like the *Journal of Longevity Research* and *Life Extension*, published by the Life Extension Foundation. They should also be attending scientific meetings like the annual meeting of the American Academy of Anti-Aging Medicine, the Gerontological Society of America, or the American Aging Association, where they can hear reports of cutting edge research and network with their colleagues.

5: *Do you see your patient as an active partner or a passive recipient?* Your role in anti-aging medicine does not stop with having a prescription filled or taking a drug. As you become more involved in your personal life extension program, you will find that you will want to read everything on the subject, subscribe to magazines yourself, perhaps start attending anti-aging medical conferences yourself. As you become more knowledgeable, your relationship with your doctor becomes more like that of a colleague. I pride myself on keeping abreast of the latest developments in my field, but sometimes it is the patient who brings a significant finding to my attention. We all learn from one another.

6:   *Do you believe in regular follow-up?* It is essential that you see your doctor every three to four months to make sure that there are no adverse side effects, to adjust the dosage if needed, and to determine if the treatment is working. Your IGF–1 levels should be tested by a reputable lab to see if they are at youthful levels. You don't want to go to a growth hormone mill where the doctor hands you some chemicals and says, "Good luck."

(In the appendix, we list the names of anti-aging specialists as well as the names of organizations that can refer you to physicians in your area who use anti-aging and advanced preventive technologies in their practice. Many of these doctors do hormonal replacement therapy, including growth hormone and growth hormone releasers.) But remember, it's your job to interview and evaluate his or her qualifications to be your personal anti-aging physician.

WHAT YOUR DOCTOR SHOULD DO

In this book, we are talking about hormones and many other substances that have powerful anti-aging and age-reversing effects. Taking these drugs is not like popping a multivitamin pill. This is serious medicine with the potential for adverse side effects, and I strongly recommend that a competent physician closely monitor their use. For this reason, I would question any doctor who prescribes hormonal therapy without first assaying the levels of that particular hormone. In the past, some doctors have been rebuked by their colleagues, and rightly so, for doing hormonal replacement therapy by the seat of their pants. This may have been excusable when hormonal assays were difficult to do and extremely costly. But today, when these assays are readily available, there is really no reason not to do them, and, indeed, omitting them may prove to be less than optimal medical practice.

Ideally your doctor should test for the levels of not only the hormone itself but its breakdown products. In the case of growth hormone, IGF–1 or somatomedin C, as it is also called, is usually tested rather than GH itself. That's because in order to test for growth hormone directly you either have to draw blood every twenty minutes over a twenty-four-hour period or do provocative testing using states or agents that induce growth hormone, like hypoglycemia (produced by giving insulin), arginine, glucagon, or L-dopa. Another way to measure overall levels of growth hormone and IGF–1 is to assay the IGF–1 binding protein 3 (IGFBP–3). The levels of this protein are dependent on growth hormone, and low levels indicate GH deficiency.

The tables below list the mean values of IGF–1, IGFBP-3, estrogen, testosterone, and DHEA at various ages. We have obtained these values from Smith, Kline and Co., which maintains laboratories throughout the country. We have not included melatonin since a good reliable and affordable assay for this does not yet exist, which unfortunately relegates melatonin replacement for the time being to the seat-of-your-pants category.

Once the target level of hormone replacement has been reached and is stable over time, the physician should continue to monitor the levels of the particular hormone with laboratory tests every four to six months on an ongoing basis, check for side effects, and adjust the dosage up or down when indicated.

HORMONAL TABLES

IGF–1 Serum (Somatomedin C)

Age	Male (NG/ML)	Female (NG/ML)
2 months–5 years	17–248	17–248
6 years–8 years	88–474	88–474
9 years–11 years	110–565	117–771
12 years–15 years	202–957	261–1096
16 years–24 years	182–780	182–780
25 years–39 years	114–492	114–492
40 years–54 years	90–360	90–360
>55 years	71–290	71–290

Insulinlike Growth Factor Binding Protein 3 (IGFBP–3)

Age	Units
30–35	1.29–4.06
35–40	1.50–3.44
40–45	1.33–3.58
45–50	1.44–2.75
50–55	1.31–2.52
55–60	1.53–2.43
60–70	1.40–3.22

Estrogen

Female	Reference Ranges
Prepubertal	12 to 57
Follicular Phase	29 to 525
Luteal Phase	126 to 478
Postmenopausal	23 to 103

Male	
Prepubertal	12 to 55
Adult	38 to 139

Progesterone

Female	nanograms per milliliter	pmol/l
Follicular phase	0.3–0.8	0.9–2.3
Luteal phase	4–20	11.6–58

Male		
	0/12–0.3	0.3–0.9

Testosterone

Males

Age	Range
20–30	280–1205
30–40	350–1010
40–50	255–1025
50–60	255–950
60–70	120–870
70–80	38–850
80–90	28–390

Dehydroepiandosterone-sulphate (DHEA-S)

Age	Male	Female
30–39	1.0–7.0	0.5–4.1
40–49	0.9–5.7	0.4–3.5
50–59	0.6–4.1	0.3–2.7
60–69	0.4–3.2	0.2–1.8
70–79	0.3–2.6	0.1–0.9

BLOOD-TESTING PROGRAM

While a program of growth hormone replacement or GH stimulation should not be undertaken without medical supervision, you can determine your own levels of many different hormones. The Life Extension Foundation offers a unique blood-testing program that can enable you to take regular blood tests without having to visit a doctor to obtain a prescription. The foundation offers all the standard blood tests as well as specialized tests for all the hormones discussed in this book.

After you order the tests you want, you receive information from the foundation to take with you to any of 10,000 facilities in the United States where you can have your blood drawn. You then ship the blood to the foundation testing laboratory, where it is analyzed. Finally, your test results and an analysis of these results are sent to you by mail. These results can then be used by you and your doctor to help determine the benefits of your hormone replacement program.

For information about this program, contact them at (719) 475–8775.

LIFE EXTENSION PHYSICAL EXAMINATION

In addition to the general workup that any physician should do, which includes blood cholesterol (LDL, HDL, and total), blood glucose, and blood pressure, a series of sophisticated tests can determine biomarkers of aging, which are useful in predicting your individual risk of age-related disease and infirmity. Dr. Vincent Giampapa, director of research, and his group at the Longevity International Institute in Montclair, New Jersey, have formulated a Biomarker Matrix Protocol that measures aging on four levels: overall body function, skin analysis, molecular analysis, and DNA analysis. Since many of these tests are expensive, you may not wish to do all of them, but we list them here as an option for the reader who wishes to pursue the ultimate in anti-aging medicine.

Level 1. Biomarkers of aging on the physiological level or overall functioning. They include:
- muscle mass/body fat ratio
- flexibility
- aerobic capacity
- bone density
- tactile response time
- forced expiratory volume
- visual and auditory tests

Level 2. Biomarkers of aging at the cellular level, which includes a skin biopsy of areas of the body not exposed to the sun, but which show tissue changes of aging in the skin. This analysis includes:
- basement membrane changes
- epidermal turnover rates
- collagen ratios
- sebaceous gland architecture
- microvascular changes
- elastic fiber content

Level 3. Biomarkers of aging on the molecular level, including biochemical assays of key biomarker hormones:
- DHEA
- HGH
- thyroid hormone
- cellular enzyme Q10
- insulin sensitivity
- heat shock proteins
- cancer oncogene blood tests
- antioxidant serum levels

Level 4. Biomarkers of aging on the chromosomal level. These are just now being developed and are more in the nature of futuristic surveys. They include telomere position and DNA strand breakage rates. The scientists at LII have also come up with a breakthrough blood test that will be able to track damage to the DNA so that you can tell exactly what effect an anti-aging regimen is having on reducing the damage to the DNA of your cells. Tests like these are the wave of the future and will transform medicine practice. By going to an anti-aging physician, you are actually participating in the medicine of the future. The results that the doctor obtains with you will become part of the knowledge base that will help propel this new science into the mainstream. In essence, we're all guinea pigs but for a good cause—the furthering of vigorous, functional, potentially limitless life span.

Replenishing the Hormones

Anti-aging hormones have captured the attention of the media and the public in recent years. First melatonin and now DHEA are being touted as the route to rejuvenation. Before that it was the replacement of estrogen for postmenopausal women and testosterone for older men. But scientific studies are revealing that many of most impressive age-reversing benefits, such as estrogen in preventing loss of bone, testosterone in building muscles, and DHEA in elevating mood may really be due to the stimulation of growth hormone by these hormones. Growth hormone may ultimately even be responsible for melatonin's "miraculous" effects, according to anti-aging researcher Greg Fahy, Ph.D.

THE BENEFITS OF MULTIHORMONAL REPLACEMENT

Replacing the major hormones that decline with age has practical value in a growth hormone enhancement program. First, replacing these hormones will make growth hormone work even better. This is true whether you need growth hormone injections or you stimulate your own endogenous levels by using the Growth Hormone Enhancement Program outlined in this section. Second, as this chapter will make clear, different hormones do different things in the body. There are reasons that all these hormones are in the body and that our bodies function at peak efficiency in our youth when our hormones are at their highest levels. Third, if you cannot stimulate your own endogenous levels of growth hormone and require injections of human growth hormone, raising the levels of the other hormones will decrease the amount of HGH you need, lowering the overall cost of treatment.

HORMONAL SYNERGY

"There is a synergy between the hormones," says Thierry Hertoghe, M.D. "If you give two, three, or four, it's like you're giving eight because of the synergistic effect. You get the best effect when you fill in all the gaps. And it allows you to use far less growth hormone, which is very expensive." By giving growth hormone as part of a multihormone replacement therapy, Hertoghe finds that he can reduce the dosage of GH in half from .5 to 2 IU a day down to .25 to 1 IU. In his own case, the thirty-nine-year-old physician has found that rectifying several hormonal deficiencies allowed him to go from 1 IU a day to .05 IU. "There is greater anti-aging efficiency and greater safety," he says.

In this chapter we will look at six hormones—estrogen, progesterone, testosterone, DHEA, melatonin, and thyroid hormone—which have been shown to have youth-restoring benefits.

ESTROGEN AND PROGESTERONE

It now appears that many of the age-reversing benefits attributed to estrogen, such as increases in protein synthesis, lean body mass, and decreased loss of bone, may actually be due to the stimulation of growth hormone. According to a 1987 article by a team of researchers headed by Michael Thorner, M.D., at the University of Virginia in Charlottesville, "Some investigators have hypothesized that restoration of pulsatile GH secretion in the elderly may reverse many of these involutional (shrinking) processes of aging. Indeed, it is possible that *the positive effect of estrogens on postmenopausal bone metabolism is in part mediated by activation of GH secretion.* [Emphasis added.]"

ESTROGEN PREVENTS DEATH

If you were born between 1900 and 1915, you may still be alive and healthy thanks to estrogen replacement therapy (ERT). In a 1996 milestone study, Dr. Bruce Ettinger compared the long-term medical outcomes of 454 women born between those years who were members of the Kaiser Permanente Medical Care Program in Oakland, California. About half the group, 232 women, used estrogen therapy for a least a year, starting in 1969, while the rest, 222 age-matched women, did not. Here's what he found among the estrogen users versus the nonusers.

- Overall mortality from all causes reduced 46 percent (fifty-three deaths for estrogen takers versus eighty-seven for nonusers).
- Coronary heart disease deaths reduced 60 percent.
- Stroke deaths reduced 70 percent.
- Cancer deaths were approximately the same in both groups, with the estrogen users having a slightly higher death rate from breast cancer and a slightly lower death rate from lung cancer.

"The overall benefit of long-term estrogen use is large and positive," the study reported, nothing that women who use this "relatively inexpensive drug can substantially reduce their overall risk of dying prematurely."

ESTROGEN'S HEALTH BENEFITS

Estrogen has many of the same benefits as GH replacement although not to the same degree. It is literally a heart saver. In studies involving thousands of women, ERT has been shown to protect against coronary heart disease and raise HDL and lower LDL levels. On the other hand, removing estrogen by taking out the ovaries does just the opposite, increasing LDL and total cholesterol and escalating the risk of heart disease, although the disease may not appear for twenty years.

Today many doctors use hormone replacement therapy (HRT), which is a combination of progesterone, or progestin (a synthetic form of the hormone) along with estrogen. This mimics the natural situation in the body prior to menopause and also cuts down the risk of endometrial cancer associated with using estrogen alone (see discussion below). While it is not clear whether adding progesterone to the postmenopausal hormonal regimen may reduce some of the benefits of estrogen on the heart, it is still more favorable for the heart than not taking any hormones.

Estrogen prevents the rapid decline in bone density after menopause, which is one of the major reasons that it is prescribed, although as noted above, this effect may be due to its stimulation of GH. In two studies, postmenopausal women who took the hormone for ten years or more reduced their rate of fracture by more than 50 percent compared with women who did not take estrogen. A 1995 study by noted estrogen researcher Annlia Paganini-Hill, Ph.D., at the University of Southern California in Los Angeles, found that elderly women between seventy-five and seventy-nine years old were 19 percent less likely to wear dentures and 36 percent less likely to have no teeth if they were on hormone replacement therapy. Some researchers suggest that the loss of teeth due to poor skeletal bone health is an early warning sign of osteoporosis.

Like growth hormone, postmenopausal HRT also has a beneficial effect on body composition. In a 1996 study of 671 women from ages sixty-five to ninety-four years, followed for fifteen years, those on hormone replacement therapy maintained a lower body mass index (the lower the BMI, the leaner the individual) than those who never used HRT. Estrogen also plumps out the skin and adds moisture and collagen, which helps to prevent wrinkles and smooths and firms the skin. The same moisturizing effect keeps the secretions flowing in the vagina, preventing or reversing the dryness and discomfort during sex as a result of menopause. And it reverses the male pattern hair growth on the face and body that some women experience.

ESTROGEN PRESERVES BRAIN FUNCTION

Menopause in many women is accompanied by mood swings, memory lapses, and difficulty concentrating. Estrogen can improve all of these, restoring clarity of thinking and a measure of serenity. In a 1994 double-blind study in women who had had their ovaries removed, psychologist Barbara Sherwin of McGill University in Montreal found that the women given estrogen supplements were able to learn and recall pairs of words more easily than the controls on placebo. Estrogen replacement in older women, according to a study by a team of researchers at the Department of Veterans Affairs Medical Center in Palo Alto, California, improved their ability to recall proper names. And in a double-blind study at the University of Southern California School of Medicine, estrogen replacement was associated with a better quality of life as shown by scores on tests of adaptation to life and depression.

Estrogen replacement not only sharpens memory and lifts the spirit, but it helps protect against Alzheimer's and shows promise as a treatment for the disease. As we reported in Chapter 12, two studies show that women on estrogen are far less likely to develop Alzheimer's. Estrogen fosters the production of acetylcholine, a neurotransmitter involved in memory, which is deficient in Alzheimer's patients. Dr. Howard Fillit, a geriatrician at Mount Sinai Medical Center in New York City, found that after only three weeks of daily treatment with estrogen, women with mild to moderate Alzheimer's disease were suddenly able to recall the day and month of the year, even though they were previously unable to do this. The women were also more alert, ate and slept better, and were more sociable.

PROGESTERONE

Estrogen and progesterone work in tandem in the body premenopausally and, increasingly, physicians believe that both hormones should be replaced postmenopausally. Progesterone by itself has a large number of benefits including promoting lipolysis (breakdown of fat), increasing energy through fat loss, protecting against endometrial and breast cancer (see below), improving mood and sexual function, and normalizing the levels of blood sugar, zinc, and copper.

ESTROGEN, PROGESTERONE, AND CANCER

Many doctors and patients have been concerned about reports that estrogen replacement raises the risk of cancer. This is particularly true of endometrial cancer, a disease that is extremely rare in premenopausal women who are still producing normal levels of both estrogen and progesterone. When progesterone is given along with estrogen for ten or more days per cycle, it not only eliminates the risk of this cancer but may actually reduce it beyond that which occurs spontaneously.

The relationship between hormone replacement therapy and breast cancer is much less clear-cut. In one well-publicized 1995 study in the *New England Journal of Medicine*, Graham Colditz, M.D., of Harvard Medical School found women who currently used estrogen replacement were 32 percent more likely to develop breast cancer, and women who used both estrogen and progestin were 41 percent more likely to develop breast cancer than women who never used hormone replacement. The risk rose in longer term users, with women on estrogen or hormone replacement therapy 46 percent more likely to develop breast cancer than nonusers. Age was even a bigger risk factor, with older women between sixty and sixty-four who had been on HRT for more than five years having 71 percent higher risk of developing breast cancer than women who had never taken estrogen.

But other studies do not support these findings. In a 1995 study reported in the *Journal of the American Medical Association*, which looked at a group of 537 patients with breast cancer compared with 492 randomly selected control women without a history of breast cancer, there was no statistical difference in the use of HRT between those who had breast cancer and those who didn't. If anything, the researchers concluded, those who used the hormone combination HRT for eight years or more had a reduced risk of breast cancer. In fact, a 1995 Australian study reveals that hormone replacement may actually have a protective effect against breast cancer recurrence. Dr. Jennifer Dew and her associates at the Royal Hospital for Women in Paddington, Australia, compared a group of 167 women who used hormone replacement therapy to relieve severe menopausal symptoms after they were treated for breast cancer with an equal number of women with a history of breast cancer who had not taken the hormone. After a follow-up period of seven years, those on HRT had a recurrence rate that was nearly half that of the control group: 9.6 percent compared with 18.5 percent. Ninety percent of the hormone users took continuous progestin along with estrogen, while only 10 percent used estrogen alone. The researchers suggest that the progestin was the good guy in the combination, first stimulating mitosis, or breast cell division (a cancerous effect), followed by apoptosis, a killing off of breast cells (anti-cancer effect). They speculate that a combined continuous regimen of estrogen and progestin could counteract the cell division needed to produce a cancer. Finally a study by Dawn Willis of the American Cancer Society, which followed more than 422,000 postmenopausal women for nine years, showed that women on estrogen therapy had a 16 percent lower risk of dying from breast cancer than those who did not take the hormone. While the study did not look at how many women got breast cancer but did not die of it, Dr. Willis suggests that the lower death rate among estrogen users may occur because they are more likely to get mammograms on a regular basis and be treated earlier.

The overall beneficial effects of combined estrogen-progesterone replacement make a good case for multihormonal replacement. By combining these hormones with DHEA and melatonin, both of which have an anti-cancer effect, and growth hormone that stimulates the natural killer cells that fight cancer, you may be able to eat your cake and have it too—that is, enjoy all the health-giving, age-reversing benefits of female sex hormone replacement while lowering the risk of cancer. (See Chapter 20 for a discussion on growth hormone and cancer.)

TAKING ESTROGEN AND PROGESTERONE

According to Dr. Hertoghe, growth hormone works best in women if it is given in conjunction with ERT. That way, he says, you need less HGH and the rejuvenation effect is more marked than if growth hormone is given alone.

The best way to replace the female sex hormones lost with menopause is to follow Mother Nature, replacing both estrogen and progesterone and using natural products rather than synthetics. Among the three major estrogens produced by the body, estradiol is most stimulating to breast tissue, estrone is second, and estriol is by far the least and is believed to protect the breast against the growth-promoting effects of the first two. Synthetic estrogen supplements are high in estrone and estradiol, while natural estrogen is high in estriol. Surprisingly, a low dose regimen of only .3 milligrams per day has the same good effect on the blood lipids as a 1.25-milligram dose, but it does not increase the risk of breast cancer even after 12.5 years of use. Natural micronized progesterone, according to a number of reports, eliminate the side effects associated with synthetic progestins, such as abnormal menstrual flow, fluid retention, nausea, depression, and weight fluctuation. Finally, if you are concerned about taking estrogen, natural progesterone alone in the form of pills or creams offers many of the same benefits, such as increased bone density, restoring libido, and elevating mood, without the associated cancer risk.

TESTOSTERONE

An article in the *New England Journal of Medicine* made headlines recently when it showed that weekly injections of 500 mL of testosterone added an average of more than one pound of lean body mass a week to male weight lifters who were pumping iron, outmuscling fellow weight lifters who received dummy injections. It also bulged the triceps and quadriceps of a second group of men who received testosterone but did not exercise. But how much of this bulking up was due to testosterone and how much to growth hormone?

Testosterone is a strong growth hormone stimulant. Several studies have found that testosterone replacement in men who are deficient raised their levels of growth hormone and IGF-1. And in 1993, a team of researchers led by Curtis J. Hobbs of the Madigan Army Medical Center in Tacoma, Washington, found that when men with normal levels of male sex hormones were given testosterone their IGF-1 levels rose from an average of 293.5 nanograms per milliliter to 354.9 nanograms per milliliter—putting them at a high-functioning level. The rise in IGF-1 levels, the authors suggest, "might serve as a mechanism for several observed effects of T [testosterone]." These include improvements in osteoporosis and body building.

THE MALE MENOPAUSE

As we mentioned in Chapter 11, men as they age go through andropause, a gradual decline in androgens, the male sex hormones. Starting in the late forties to early fifties, the male hormones are about a third to a half of what they were in young manhood. By the age of eighty to ninety, testosterone has decreased by about 60 percent. Both total serum testosterone and "free" testosterone, the part that is not bound to protein and is believed to be the active component, decline with age. The loss of testosterone, like the loss of growth hormone, contributes to the familiar "pot belly" and declining muscle tone in middle-aged men (see Chapter 9).

While menopause announces itself clearly in all women with the cessation of monthly periods, andropause, the male equivalent, is mostly silent, with a gradual decline in male hormones that varies greatly in degree from one man to another and can occur anywhere from age thirty-one to age eighty. The symptoms of the male menopause were studied by Danish researcher Dr. Jens Moller, who found that the large majority of men with low testosterone levels had a reduction in libido, difficulty in having and maintaining an erection, and decreased sexual satisfaction. Other symptoms included fatigue, depression, irritability, aches and pains, and stiffness.

THE MISUNDERSTOOD HORMONE

For a long time it was believed that high testosterone levels in men promoted heart disease and was the reason that men got heart disease at a younger age than women. But a new study at Columbia University College of Physicians and Surgeons suggests the male sex hormone is good for the heart. Gerald Phillips, M.D., professor of medicine at the university, found that of fifty-five men undergoing X-ray exams of their coronary arteries, those with higher levels of testosterone had high levels of protective HDL cholesterol, while those with low testosterone had higher degrees of heart disease as shown by their coronary clogging. Other research shows that testosterone supplementation may lower harmful LDL cholesterol as well. In one study, men given testosterone replacement had 9 to 11 percent decreases in total cholesterol. "A low testosterone level may lead to atherosclerosis," according to Phillips, who reports "that testosterone may protect against atherosclerosis in men through an effect on lipoprotein HDL. Administration of testosterone to men has been reported to decrease risk factors for heart attack. Low testosterone is also correlated with hypertension, obesity, and increased waist-to-hip ratio"—all heart attack risk factors.

TESTOSTERONE REPLACEMENT

Equal rights for men is on its way. Because the male menopause was not taken seriously until very recently, there have been few double-blind clinical trials of testosterone replacement therapy (TRT), compared with ERT in women. The studies that have been done tend to bear out that TRT is as potent in its anti-aging effect as estrogen and progesterone replacement in women. It renewed strength, improved balance, raised red blood cell count, increased libido, and lowered LDL cholesterol. In one double-blind study, twelve of thirteen men who were receiving testosterone rather than placebo could tell they were on the active drug because they felt more aggressive and energetic at work. In addition, they reported having better sexual performance, initiation of sexual intercourse, and increased ability to maintain an erection.

Testosterone also saves bone and helps prevent osteoporosis in men. While most people tend to think of bone loss as a problem in postmenopausal women, men after age sixty have a dramatic rise in hip fractures, with the rate doubling every decade. Hypogonadal males, or those with abnormally low levels of testosterone, are six times more likely to break a hip during a fall than those with normal testosterone levels, according to Dr. Fran Kaiser, associate director of geriatric medicine at St. Louis University of Medicine. In studies of TRT in both young hypogonadal men and older men with low testosterone levels, there have been increases in bone density, bone formation, and bone minerals, such as osteocalcin.

Testosterone also plays a role in immune function, although not always in the desired direction. In animal studies castration that cuts off the supply of testosterone improved many immune responses, while replacing testosterone caused a decline in immune response. But it is protective against autoimmune disease, which affects older people. Androgen treatment in men improved autoimmune conditions, such as rheumatoid arthritis or systemic lupus erythematosus. Testosterone, like estrogen, also heightens mood and sense of well-being and increases some mental functions, particularly visual spatial ability used, for example, in reading a map.

While testosterone has yet to enjoy the popularity with older men that estrogen has with women, some physicians are starting to use it in their practice on a regular basis. One of these is Stanley

Korenman, M.D., professor of endocrinology at UCLA. "My basic position is that old men have no reason to have less testosterone in their circulation than young men," he says. "Therefore, if someone is hypogonadal compared to young men, I tend to give them a three-month trial with testosterone and see how much better they do. A substantial proportion of them really feel better, are more active, aggressive, and interested in what's going on. Their mood is improved and their sexual function and interest is improved, although it doesn't improve impotence." The two side effects to watch out for, he says, are higher levels of PSA, the Prostate Specific Antigen, a test for which is used in the detection of prostate cancer (see discussion below) and a rise in hematocrit, a measure of blood volume. "The most important complication is an increase in hematocrit. You have to make sure that the blood continues to flow like cranberry juice and does not get too thick." He has not seen much change in PSA levels with testosterone replacement.

TESTOSTERONE AND CANCER

In the same way that women worry about estrogen increasing the incidence of breast cancer, men are concerned about the possible risks of prostate cancer with testosterone. Prostate cancer is the second most diagnosed malignancy in men and the second leading killer of men after heart disease. The prostate gland is a small, chestnut-sized organ that sits just below the urinary bladder and has among its functions the production of semen and storage of sperm. Signs of prostate abnormality, which can occur after age forty, include frequent daytime and nighttime urination, slight pain or burning sensation during urination, dribbling or stopping urine flow, and leakage of urine. The best way to detect prostate cancer in the early stages is by testing PSA levels in addition to the digital rectal exam (DRE), where the physician palpates the back of the prostate with a finger. Because testosterone has the potential to stimulate prostate cells, there is concern that it could aggravate problems like benign prostatic hyperplasia, a noncancerous enlargement of the prostate, or promote an undetected cancer. For this reason, it is important to have regular prostate exams and PSA tests. Dr. Michael Perring, medical director of the Optimal Clinic in London, who has treated over 800 patients with symptoms of andropause, recommends three yearly ultrasounds in addition to a PSA test every six months.

PROTECTING THE PROSTATE WITH HORMONES

Just as with estrogen and breast cancer, it may be that multihormonal replacement with testosterone, DHEA, melatonin, and growth hormone offers the best defense for men against prostate cancer. Chein and Terry believe that total hormone replacement including melatonin and DHEA (see below) has immune rejuvenating and cancer-surveillance properties that protect against prostate cancer. There has been no rise in PSA levels or reported cases of cancer in over 800 patients that they have treated. In one patient who had extremely high PSA levels indicating prostate cancer, which was confirmed on biopsy, the PSA levels dropped to near normal after he was treated with growth hormone! (See Chapter 3.)

TAKING TESTOSTERONE

The same risk-versus-benefit considerations apply to hormone replacement in men as they do in women. Among the reported adverse side effects are atrophying of the testicles with long-term use, high red blood cell and hematocrit count, depression, fluid retention, reduced sperm count and volume of semen, and reduced HDL cholesterol. Testosterone is a potent medication and should only be used if your levels are below normal.

The best way to maintain normal levels of free circulating testosterone (and growth hormone) is with regular vigorous exercise, particularly treadmill running and weight-training.

If your levels of free testosterone are low and you wish to bring your levels up to that of a young man, use pure natural testosterone, which may be administered by intramuscular injections, suppositories, a patch attached to the scrotum, a cream applied to the scrotum, or oral micronized capsules or sublingual lozenges. Pellets implanted under the skin are now being used experimentally. In Europe a percutaneous gel, which crosses the skin into the bloodstream, closely mimics the way your own testosterone works in the body. This year the Androdrem transdermal testosterone patch came onto the market. This patch, which can be worn anywhere on the body, allows a continuous release of testosterone and appears to have the benefit of sustained blood levels without the need of injections. Avoid oral testosterone, which is the least effective, and synthetic testosterones, such as methyl testosterone, which have been associated with liver toxicity.

TESTOSTERONE IN WOMEN

Testosterone replacement is becoming increasingly popular among women who claim that it makes them feel sexier and energized. Since women's ovaries and adrenals do make a small amount of testosterone, which is cut in half during menopause, replacing it seems to help some women who continue to have postmenopausal symptoms despite estrogen replacement. According to Korenman, low dose replacement rarely causes problems, but higher doses can result in masculinization, such as unwanted hair growth and deep voice, increased blood pressure, and risk of heart disease.

DHEA

By now, most people are familiar with the all the hoopla surrounding the hormone with the tongue-twister name, dehydroepiandrosterone, better known as DHEA. It has been called the "fountain of youth" hormone and touted as the ultimate weight-loss, muscle-building, cancer-preventing, feel-good drug. But here, again, many of its benefits stem from the fact that it that it raises the level of IGF–1 and growth hormone.

WHAT IS DHEA?

The manufacture of DHEA starts in the pituitary axis, which releases ACTH (adrenocorticotrophic hormone). ACTH signals the adrenal glands to manufacture DHEA from cholesterol. It is released into bloodstream as DHEAS (DHEA sulfate). Dubbed the "mother of all steroids," DHEA is the most abundant steroid in the human body and is involved in the manufacture of testosterone, estrogen, progesterone, and corticosterone. The decline of DHEA with age parallels that of growth hormone, so that by age sixty-five, your body makes only 10 to 20 percent what it did in at age twenty.

Many anti-aging specialists believe that replacing DHEA in later life will have multiple beneficial effects even in people who are ostensibly healthy.

BENEFITS OF DHEA

What does DHEA do? You might as well ask what doesn't it do. It is the most abundant steroid hormone in humans, made primarily by the adrenal cortex, but also synthesized by the brain and skin. In animal studies it is anti-obesity, anti-diabetes, anti-cancer, anti–autoimmune disease, anti–heart disease, anti-stress, anti–infectious disease, in other words, an all-around anti-aging drug. It extends the life of laboratory animals by 50 percent, and mice given the hormone look younger and better, maintaining the glossiness and coat color of their youth. It may have a life-extending effect in men as well, although not as great as had been originally reported in a study that now spans nearly nineteen years. While Elizabeth Barrett-Connor and associates at the University of California in San Diego in 1987 reported a 70 percent drop in mortality from heart disease in men with high DHEAS levels, a 1995 follow-up study of the same group found only a 20 percent drop in deaths when compared with those who had low DHEAS levels. Higher DHEAS levels did not protect women, who had a slightly higher, though statistically nonsignificant, risk of dying from cardiovascular disease. But in another study of the oldest old over age eighty-five, the one-year survival rate was lower among men who had higher DHEAS, while women with high DHEAS levels tended to survive longer than other women.

DHEA appears to have terrific potential as an immune system enhancer. Dr. Raymond Daynes, head of the division of cell biology and immunology at the University of Utah at Salt Lake City, found that it rejuvenated many parameters of immune function in mice, including the proliferation of T-cells and IL cytokine 2, which decline with age. And a study of mice with viral encephalitis found that DHEA eased some symptoms, reduced the death rate, and postponed both the onset of the disease and death. Old people do not respond as well to vaccines as young people. But when Daynes gave old mice vaccines laced with DHEA, their ability to mount defenses to such diseases as hepatitis B, influenza, diphtheria, and tetanus equaled that of a young animal. The animals he placed on

DHEA replacement therapy, according to Daynes, also looked "far, far healthier in their later months."

In humans, low levels of DHEA have been shown to predict heart disease in men and breast cancer and ovarian cancer in women. It may play a role in maintaining brain cells and protecting against Alzheimer's disease. Brain tissue contains five to six times more DHEA than any other tissue in the body. A study of sixty-one men, ages fifty-seven to 104, who were confined to a nursing home found that the lower the DHEA level, the more dependent the person was and the more difficulty he had in carrying out daily activities. It also has anti-diabetic action, increasing insulin sensitivity in mice and actually preventing the disease in rats bred to develop diabetes. Some clinicians report that their patients need less insulin when they are on DHEA.

One very exciting aspect of the hormone is its potential as an anti-obesity drug. In one study, high doses of DHEA (1,600 milligrams per day) given for four weeks, caused a 31 percent decrease in body fat in four of five subjects with no overall weight change, implying a substantial increase in muscle mass. Their LDL levels also fell by 7.5 percent, showing that DHEA protected their hearts as well. The weight loss possibilities of the hormone have been explored by Arthur Schwartz, Ph.D., of the Fells Institute for Cancer Research and Molecular Biology at Temple University, a pioneer in DHEA research. He found that animals on the steroid lost weight regardless of how much they ate! When he took normal-weight mice and controlled their activity and diet for four weeks, they reacted exactly like the humans, with four out of five losing 31 percent fat while their overall weight remained the same.

There are a number of explanations for the anti-obesity effect. First, DHEA inhibits an enzyme called glucose–6-dehydrogenase (G6DPH), which may block the body's ability to store and produce fat. Second, Schwartz found that it stimulates cholecystokinin (CCK), which signals the body to feel full. Third, it may work through IGF–1 to shift the metabolism from producing fat to creating muscle and energy.

ANTI-AGING TEST

In the first real anti-aging test in people, DHEA passed with flying colors. But it appears that the hormone accomplished its results the same way that growth hormone does—by raising the levels of IGF–1. The 1994 double-blind, placebo-controlled, cross-over study of DHEA in aging men and women was carried out by Dr. Arlene Morales, Samuel Yen, and their associates at the University of California School of Medicine in San Diego. It involved seventy-one women and thirteen men between the ages of forty and seventy who were on 50 milligrams of DHEA for three months and a placebo for three months. When they were on the drug, their levels of DHEA and DHEAS rose to those of a young adult. Although the subjects did not know when they were on a dummy pill or the real thing, 82 percent of the women and 67 percent of the men reported an improved sense of well-being, which included such aspects as better quality of sleep, increased energy, improved ability to handle stress, and feeling more relaxed. Five of the volunteers also noted improvement in chronic joint pain and mobility. There were no changes in body composition, even though other studies, including a year-long one by Dr. Yen, found that DHEA increased muscle mass in men and women and that men gained strength and lost fat.

Most intriguing, the San Diego group found that DHEA caused a significant rise in IGF–1, although it did not affect the twenty-four-hour measurement of growth hormone levels. They speculate that restoring the levels of DHEA may stimulate the liver to produce more IGF–1 or generate more growth hormone receptors. In other words, as with estrogen and testosterone, much of the anti-aging benefits attributed to DHEA may actually be due to the stimulation of the GH and IGF–1 system!

In a later report on that and other clinical studies carried out by the San Diego group, Yen told a three-day conference on DHEA and Aging sponsored by the New York Academy of Sciences that "because the GH-IGF–1 system and immune function decline with aging in parallel to DHEA, we hypothesize that these concomitant changes may be functionally linked." The age changes that he was referring to included reduced protein synthesis, decreased lean body mass and bone mass, and increased body fat. When he and his associates gave replacement doses of 50 milligrams per day to both men and women, they found a twofold rise in male hormones in women and only a small rise in men. "This [rise] was associated with a remarkable increase in perceived physical and psychological well-being for both men (67%) and women (84%) and no change in libido," they reported. There were no side effects. They also found that daily doses of 50 milligrams and 100 milligrams increased knee extension flexor strength in men, but not in women. Fat mass decreased in men although not in women. The 100-milligram dose, however, proved to be too high in women since it resulted in three- to four-fold rise in male hormones in women, with one woman developing facial hair that went away at the end of the study.

A one-year study of 100 milligrams of DHEA in men and women between the ages of fifty and sixty-five, "confirmed the abil-ity of DHEA to induce an increase in IGF–1," according to Yen and his associates. "Furthermore, biologic end points of increases in lean body mass and muscle strength of the knee were observed."

They also believe that the immune enhancement of DHEA is due to IGF–1 stimulation. The immune benefits that have been reported in older people include increased natural killer cell activity and improvement in the autoimmune disease systemic lupus erthy-matosus, with alleviation of symptoms and reduced need for corti-costeroids. "The mechanisms by which DHEA exerts its lym-photrophic effects is unknown," they write. "The temporal syn-chrony of the increase in circulating IGF–1 and immune activation by DHEA suggests that the immunoenhancing effects of DHEA may be mediated by IGF–1 by virtue of its immune regulating prop-erties which have been demonstrated both in vivo and in vitro."

TAKING DHEA

Some people are now using DHEA as an inexpensive alternative to growth hormone. It costs as little as $30 for a one-month supply of 100-milligram capsules. It has not been shown to have as dramatic an age-reversal effect as GH and it has yet to be proven in long-term human studies. It is a food additive, but the FDA has approved it by prescription only, although it is freely available at most health food stores and drugstores. Since it can metabolize to estrogen and testos-terone in the blood, the same precautions apply as for those hor-mones. It should not be taken by anyone with a history of prostate or ovarian cancer. Dosages commonly range between 25 and 150 milligrams, but it is best to start at the low-dose end of about 25 to 50 milligrams per day and raise the dosage later if needed. In Yen's study, 50 milligrams was enough to bring DHEA up to youthful levels. For best results, take in divided doses three or four times a day. Have your DHEA levels measured every two to three months. Since taking too much DHEA can turn off your own internal production, you can try taking it every other day and alternating with a DHEA precursor, such as dioscorea, or wild yam capsules.

MELATONIN

This is another of the wunderkind hormones that has captured the public imagination. Even people who are not into anti-aging are popping it for sleeping and jet lag. It is one of the most potent antioxidants yet discovered, an immune booster, cancer fighter, heart helper, potential AIDS therapy, and mood elevator. Greg Fahy, Ph.D., of the Naval Medical Research Institute in Bethesda, Maryland, speculates that growth hormone may hold the key to why melatonin supplements have such a widespread anti-aging effect. "This remains theory," he says, "but the fall in melatonin in the brain triggers puberty. As you go through puberty, your growth hormone levels start to fall. So it stands to reason that these things are connected. It may be that the reason that melatonin is good is that by putting it back you can stimulate GH release." In fact, so close is the relationship between the two hormones that a prominent researcher in the field, Russell Reiter, Ph.D., advises taking melatonin as a cheap and easy way to increase growth hormone. We disagree. While melatonin in low doses is an important part of multihormonal replacement, it can no more replace growth hormone than can DHEA. Only growth hormone replacement either by injection or with the Growth Hormone Enhancement Program can bring the IGF–1 back to the youthful levels required to reverse the ravages of the age on the mind and body.

THE AGING CLOCK

Like a vampire, melatonin comes alive in the dark and shrinks in the light. It is secreted by the pineal gland, a small organ set behind and between the eyes, which is often referred to as the "third eye" that sees more deeply and truly than the other two. The pineal gland is the timekeeper of the brain, helping to govern our circadian or twenty-four-hour daily rhythms, such as the sleep-wake cycle, as well as seasonal rhythms, such as migration, mating, and hibernation in animals. In a bold series of experiments, Italian immunologist Walter Pierpaoli, M.D., Ph.D., and Russian scientist Vladimir Lesnikov transplanted the pineal gland from old mice to young mice and found that they were able to speed up aging. When they performed the reverse experiment, transplanting the pineal from young mice to old, it gave the mice a whole new lease on life—allowing them to regain youthful function and maintain it throughout their maximum life span. Pierpaoli and other colleagues also extended the life span of mice by as much as 25 percent with melatonin supplementation. The treated mice also appeared younger, healthier, and more vigorous. The rejuvenation also extended to their sexual function, not only reawakening youthful interest and vigor, but actually repairing and regenerating their sexual organs so that they were comparable to those of younger animals.

MELATONIN AND DISEASE

There are no clinical trials of melatonin in aging people or any long-term, double-blind studies. But research in humans shows that it boosts the immune system, lowers cholesterol in people with high levels, and shows promise in preventing and treating cancer. It is also a highly effective and safe remedy for inducing sleep and overcoming jet lag. In a 1995 study, melatonin strengthened the immune system, particularly those aspects that fight cancer, in twenty-three cancer patients undergoing chemotherapy. In addition to stimulating immunity directly, the hormone acts as a buffer against stress, counteracting the suppressive effects of stress hormones on the immune system. Melatonin also significantly increased the one-year survival rates of people with metastatic lung cancer in a 1992 study by Dr. Paolo Lissoni and his colleagues at San Gerardo Hospital in Monza, Italy. In another study by Lissoni of 200 cancer patients with advanced solid tumors with a life expectancy of six months, a combination of melatonin and immunotherapy caused complete tumor regression in 2 percent of the patients, caused partial regression in 18 percent, and stabilized the disease in 38 percent. Melatonin may play a role in preventing cancers of the breast and prostate as shown by the fact that patients with these diseases have very low levels of this hormone.

MELATONIN AND FREE-RADICALS

The constant bombardment of free-radicals from the breakdown products of food and oxygen is a major contributor to aging and the diseases of aging, such as heart disease, cancer, and autoimmune diseases. Melatonin is one of the most effective free-radical scavengers yet discovered, capable of penetrating every cell of the body and working on both the outside of the cell—the lipid-rich membrane— and the water-filled inside. Dr. Russel Reiter and his colleagues at the University of Texas Health Science Center in San Antonio gave melatonin to rats before feeding them food laced with carcinogens that are known to produce DNA-damaging free-radicals. The melatonin-treated rats had 41 to 99 percent less genetic damage than untreated controls. The more melatonin the rats were given, the greater their protection from damage.

MELATONIN, SLEEP, AND JET LAG

Melatonin may be the closest thing we have to a natural sleeping pill. Like GH, its release is almost entirely confined to the nighttime. The amount of melatonin we make at night is directly related to how well we sleep. Elderly people with insomnia have half the amount of melatonin as young people. The declining levels of growth hormone in the aged also contribute to sleep problems. Dr. Steve Novil, a researcher at the American Longevity Research Institute in Chicago, has been treating patients with melatonin for six years. One eighty-year-old woman patient had a problem common to many older people—inability to return to sleep after getting up to urinate several times a night. Within a week of treatment with 3 milligrams an hour before bedtime, she awoke to urinate less frequently and when she did, she fell right back to sleep. In addition to better quality of sleep, she reported having vivid, pleasant dreams for the first time in years and woke up feeling refreshed. Studies found that low doses of melatonin (.3 milligram and 1.0 milligram) reduced the time to fall asleep, increased actual sleep time, reduced the number of awakenings, and increased the quality of sleep as reflected in deep or slow wave sleep. Since the greatest bursts of growth hormone are produced in slow wave sleep, insomniacs on melatonin may be triggering GH releases, although little on this has been published to date. In the same way that melatonin resets the body clock at night so that you sleep better, it can reschedule the body after shift work or zipping across time zones throws the biological rhythms out of whack. A British endocrinologist who tested melatonin on some 400 travelers found that it reduced the symptoms associated with jet lag, like fatigue and disorientation, by 50 percent. French researchers conducted a double-blind study in which thirty subjects, who had had problems with jet lag in the past, flew from the United States to France. Those who had been given 30 milligrams of melatonin on the day of the flight and for three days thereafter were able to sleep and focus better and experienced fewer mood swings than those who received a dummy pill.

TAKING MELATONIN

Unfortunately, finding the right dose of melatonin is a matter of trial and error. Some people do well on as little as 200 micrograms, while others seem to require as much as 60 milligrams. Unless you need it to sleep well or to reset your body clock after crossing time zones, there does not appear to be much need for anti-aging purpose until you are sixty or older. To keep your pineal gland in a good melatonin-producing state before that, do what human beings did for aeons before Edison lighted up the world: Sleep in a darkened room and roll up your blinds first thing in the morning and soak up the sunlight. Remember the vampire effect.

If you are over sixty, start out by taking low doses and only before bedtime. Taking doses that are far beyond physiologic, or replacement, doses may turn down your own natural production of the hormone. To find the dose that works for you, start with the lowest amount, about .5 to 1 milligram. about a half hour to two hours before bedtime. Most people feel drowsy within thirty minutes. Faster-acting lozenges or sublingual tablets are also available. With the proper dosage, you should be able to fall asleep and sleep better. If you feel groggy or headachy in the morning, reduce the dosage in increments of .5 milligram until you find the amount that's right for you. If you have trouble sleeping, try raising the dosage to 5 or 10 milligrams, although some people require 20 milligrams or more to get an effect. Many researchers prudently recommend taking it every other day or even less often so that you don't reduce the capacity of your pineal gland to secrete its own stores of melatonin. If you decide to go off melatonin, taper off by cutting the dosage back in increments over several days for a period of one to two weeks to avoid rebound insomnia.

Cautions

Melatonin should not be used for *any* reason by pregnant or nursing mothers, children, women trying to conceive (high doses act as a contraceptive), or people who are on prescription steroids or who have mental illness, depression, severe allergies, autoimmune diseases (such as multiple sclerosis), or immune system cancers (such as lymphoma and leukemia). Check with your personal anti-aging physician before beginning any hormonal replacement therapy.

THYROID HORMONE

The thyroid, a small, butterfly-shaped gland located behind the hollow of your throat, affects virtually all your metabolic processes, including body temperature and heartbeat. According to Karlis Ullis, M.D., director of the Sports Medicine and Preventive Medical Group in Santa Monica, California, it is important to rule out thyroid hormone deficiency before starting a growth hormone program, because GH increases the metabolism, which will use up thyroid hormone more quickly, creating an even greater deficiency. But it also works the other way around. Jens Sandahl Christiansen of the Kommunehospital in Aarhus, Denmark, who works with growth-hormone deficient patients, says lack of growth hormone can cause hypothyroidism. When GH is replaced, it often normalizes the thyroid.

SYMPTOMS OF THYROID DEFICIENCY

Hypothyroidism—an insufficient production or absorption of thyroid hormone (TH)—can look a lot like aging. A primary symptom is a general lack of energy with weakness and moving too slowly.

Other symptoms of thyroid deficiency may include being susceptible to colds, viruses, and respiratory ailments; heavy labored breathing; muscle cramps; persistent low back pain; bruising easily;

mental sluggishness; emotional instability with crying jags, mood swings, and temper tantrums; getting cold easily, particularly in the hands and feet; dry, coarse, leathery or pale skin; coarse hair or loss of hair; loss of appetite; stiff joints; and atherosclerosis.

While thyroid hormone doesn't fall in a linear way with age the way DHEA or growth hormone does or shut off the way estrogens do in women, some studies indicate that as many as 15 percent of people over age sixty have subclinical hypothyroidism. Stephen E. Langer, M.D., author of *Solved: The Riddle of Illness*, estimates that 40 percent of the population may suffer from deficient thyroid function in some way.

THYROID AND THE DISEASES OF AGING

Low TH levels may not only make you feel old, they can propel you into the diseases and conditions associated with aging. Since TH governs your metabolism, low levels can interfere with your ability to keep your weight down. The changes in body composition that we discussed in Chapter 9 can also be due to lack of thyroid hormone and, conversely, decrease in lean body mass can interfere with the activity of an enzyme that converts thyroxine (T4) into tri-iodothyronine (T3), which is five times more potent than thyroxine. Adequate levels of TH are also needed to regulate blood sugar, so that a deficiency of this hormone may lead to mature-onset diabetes. Low TH can also greatly increase the risk of heart disease by raising the levels of cholesterol and triglycerides and may cause high blood pressure. Finally, a number of studies have linked low iodine levels to increased risk of cancer. Iodine is essential to the function of the thyroid and lack of it in the food or water can cause goiter, an enlargement of the thyroid gland. Dr. J.G.C. Spencer of Frenahay Hospital in Bristol, England, found that "goiter belts" in fifteen countries across four continents had higher than average cancer rates. And according to Dr. Bernard Eskin, director of endocrinology at the Medical College of Pennsylvania, iodine deficiencies are associated with breast cancer in both humans and animals. There have also been reports in the *Neurology Journal* and the *Southern Medical Journal* of cases in which thyroid supplements reversed certain types of cancer.

SELF-DETECTION OF HYPOTHYROIDISM

You can determine if you are low in thyroid by taking the Barnes Basal Temperature Test.

1. Place a shaken down thermometer next to your bed before going to sleep.
2. As soon as you are awake, place the thermometer under your armpit and leave it there for ten minutes before getting up.
3. Record the temperature. If it is below normal rising temperature (97.8 to 98.2°F) for two consecutive days, you are very likely to be hypothyroid. (Menstruating women should wait until after the first day of their period before taking this test.)

If positive, you should double-check your own findings by having your doctor do the new TSH (thyroid stimulating hormone) test. Unlike the old test, which required measuring uptake of radioactive iodine which you ingested, this is a simple blood test in which the TSH is measured by radioimmunoassay. Many doctors believe that the TSH test provides the best indication of deficient thyroxine production, or hypothyroidism.

Do not attempt to treat yourself with thyroid supplements if your home test indicates hypothyroidism. Too much thyroid hormone is just as bad for the health as too little. Hormones are serious medicine and their replacement should be done under the supervision of a physician, preferably one involved in anti-aging medicine.

As you will see in the succeeding chapters, most of the so-called anti-aging drugs and supplements, as well as life-extending practices, such as vigorous exercise and caloric restriction, have been shown to selectively increase growth hormone. The fact that growth hormone is induced by an abundance of natural substances and daily activities from sleep to stress indicates its central importance to maintaining bodily function. More and more, it is beginning to look as if growth hormone is *the* hormone of youth.

Now let us show you how you can induce your pituitary gland to release its own stores of human growth hormone.

Growth Hormone Naturally: A Guide to GH-Releasing Nutrients and Drugs

The heart of the Growth Hormone Enhancement Program is the use of GH releasers or agonists, substances that induce the release of growth hormone from the pituitary. Using these agonists, you can jog your pituitary to secrete extra GH, which you can then maximize to full potential using the Growth Hormone Enhancement Program of dietary manipulation and GH-releasing exercises contained in Chapters 17 and 18. According to Mauro Di Pasquale, M.D., a world-class power lifter and one of the most knowledgeable experts in the area of anabolic and growth hormone–releasing compounds, "You don't have to increase growth hormone very much to get a 10 to 20% rise in IGF-1 levels, which can have a definite effect on the body."

The following is an informational guide to the natural substances and drugs that have been shown to increase levels of GH and IGF–1, including several new products that have been specially formulated for the purpose of raising GH level. Many of these are available from your local health food store or from suppliers listed in the appendix. Others require a doctor's prescription. If you are using GH-releasers, be sure to take a high potency daily vitamin, antioxidant complex, and mineral supplement so that the nutrients the body needs to build cells and tissue are on board. While suppliers for all these products are listed in the appendix, you should not undertake self-treatment without medical supervision. Many of these products can have adverse side effects or contraindications that make them dangerous for certain users.

AMINO ACIDS

Twenty amino acids form the building blocks of all proteins and, as such, are needed for the body to make the proteins of enzymes, many hormones, muscle, bone, skin, organs, etc. A number of the amino acids have been shown to induce growth hormone secretion. For most people, supplementation with amino acids will be sufficient to stimulate elevated levels of GH. They have the advantage of being cheap—less than $1 a day—easy to take, safe, generally free of side effects, and available from your corner drugstore or health food store.

Arginine: The GH-Provocateur

What it is
An essential amino acid, meaning that the body cannot create amino acid on its own but must get it from the foods we eat.

Effects on GH
There is no doubt that arginine causes the secretion of growth hormone. A 15- to 30-gram intravenous infusion of arginine is used as a standard endocrinological test to provoke the pituitary into releasing growth hormone (see Chapter 14). Dirk Pearson and Sandy Shaw recommended arginine and ornithine as GH-releasers in their first book, *Life Extension: A Practical Scientific Approach*, turning these amino acids into best-selling nutrients that disappeared from the shelves quicker than cranberry sauce on Thanksgiving. Shaw took 10 milligrams of arginine a day on an empty stomach as a GH-releaser to speed healing after breaking her foot. About forty-five minutes to one hour after taking the arginine, she did three

minutes of bench presses while lying on her back. With that regime, she lost 25 pounds of fat and put on five pounds of muscle in six weeks.

A number of clinical studies on various doses of arginine and arginine in combination with lysine (see below) have shown widely varying effects on growth hormone from no effect to a dramatic synergistic increase in GH. In one 1980 study by Mathieni, even 200 milligrams was enough to elicit a significant rise in GH release. Another study at Kent State University in Ohio showed a diminishing response to the growth hormone among those who were thirty to thirty-four years old compared with those eighteen to twenty-one. Within these age groups, those with low body fat and high aerobic capacity had the highest GH responses. The doses used in the study were .04 grams per kilogram of the subject's weight, .16 grams per kilogram, and .28 grams per kilogram, or about 3 grams, 12 grams, and 21 grams, respectively, for a 165-pound person. The most effective dose was the middle dose, with the highest dose causing diarrhea and producing the lowest growth hormone response.

It keeps working even into old age. A study at the University of Turin, Italy, showed that even though people in their seventies had lower response than either children or young adults to arginine, the nutrient still boosted their blood levels of GH to triple the average for their age group!

Arginine also helps to improve exercise performance, because it is one of the main ingredients, along with glycine, that the liver uses to make creatine. Supplements of creatine monohydrate are very hot now in the bodybuilding community because they raise the level of high-energy creatine phosphates within the muscle and nerve cells needed for high-intensity, short-duration exercises. So with arginine you get more bang for the buck—higher growth hormone levels and the raw material for your cellular batteries. (See metabolic enhancers in Chapter 18.)

How it works

Arginine appears to stimulate GH by blocking the secretion of the growth-hormone inhibitor somatostatin. It also greatly enhances the effect of growth hormone–releasing hormone when they are given together.

Anti-aging benefits

Positive claims for arginine include increasing fat burning and building muscle tissue probably through the stimulation of growth hormone, increasing the weight and activity of the thymus gland, boosting immunity, fighting cancer, promoting healing of burns and other wounds, protecting the liver and detoxifying harmful substances, and enhancing male fertility (almost all of which are enhanced by GH). It also restores sexual function in impotent men. In a 1994 study by Drs. A.W. Zorgniotti and E.F. Lizza of the department of urology/surgery at New York University School of Medicine, six of fifteen men who took 2,800 milligrams of arginine a day for two weeks had renewed sexual performance, specifically improved erection, yet none of the men on the placebo did. The researchers believe that arginine worked because it is a precursor of nitric oxide, which plays a key role in initiating and maintaining an erection.

Clinical usage

Arginine supplements should be effective in raising growth hormone levels, especially in people under fifty. You can also take it as a stack (bodybuilder lingo for combination of drugs or nutrients) with other amino acids, such as ornithine, lysine, and glutamine (see below).

Dosage

2 to 5 grams on an empty stomach one hour before exercise and before sleeping. (See "A Note on Stacking the Amino Acids" below.)

Side effects

The dosage required to markedly stimulate GH may cause stomach upset and nausea on an empty stomach, which is the way it should be taken. This can be minimized or eliminated by starting at a low dosage of 1 gram and building up slowly.

Ornithine—Son of Arginine

What it is

A nonessential amino acid, meaning the body can synthesize it from other food sources. Ornithine can be made from arginine and is very similar in structure.

Effects on GH

According to Pearson and Shaw, on a per gram basis, ornithine is about double the effectiveness—and cost—of arginine. They report that 5 to 10 grams of L-arginine and 2.5 to 5 grams of ornithine taken on an empty stomach at bedtime cause GH release. Studies of ornithine have shown the same variation in individual response as occurs with arginine.

Anti-aging benefits

Shares many of the properties of arginine, but it has also been shown to regenerate the liver in animals.

Clinical usage

Best used as a stack with other amino acids like arginine, lysine, or glutamine.

Dosage

2 to 5 grams at bedtime. (See "A Note on Stacking the Amino Acids" below.)

Side effects

As with arginine, high doses of about 170 milligrams per kilogram caused runny, watery diarrhea in eleven out of twelve subjects.

Lysine: Arginine Booster

What it is

An essential amino acid that affects bone formation, height, and genital function.

Effects on GH

A 1981 study by Italian researcher A. Isidori, M.D., and his associates at the University of Rome found that the combination of 1,200 milligrams of lysine and 1,200 milligrams of arginine in fifteen male volunteers between the ages of fifteen and twenty was ten times more effective than taking arginine alone. According to the researchers, "we could demonstrate that the association of the two amino acids does result in the release of biologically active hormone able to affect peripheral cellular receptors and thus cell growth in general." The fact that lysine and arginine together were active in oral form, say the researchers, "is clearly of considerable importance in clinical and diagnostic practice, where it offers a more practical and physiological approach."

While the combined amino acids may work in younger people, Emiliano Corpas and his group at the Gerontology Research Center in Baltimore found that in men over age sixty-five, arginine/lysine even in doses more than twice that of the Italian study did not raise either growth hormone levels or IGF–1. Larger doses beyond the 6 grams of arginine and 6 grams of lysine used in the study may be effective, say the researchers, but are associated with diarrhea and other adverse gastrointestinal effects.

Anti-aging benefits

According to Roy Walford, there is evidence that a combination of lysine and arginine may increase thymic hormone secretion in older animals and humans, partially reversing the immunodeficiency of aging. Again this could be GH-related. It also effectively reduced the recurrence of herpes simplex infections at dosages of 1.25 grams in a 1984 Mayo Clinic study.

Clinical use

The smart money says to stack lysine with arginine, ornithine, and/or glutamine for the best effect. (See below.)

Dosage

1 gram on an empty stomach, 1 hour before exercise and before sleeping.

Side effects

None at suggested doses. Well tolerated with low toxicity. Diarrhea and other adverse gastrointestinal effects have been seen at high doses. (See "A Note on Stacking the Amino Acids" below.)

Glutamine: New Kid on the Block

What it is

The most abundant amino acid in the body. It is a conditionally essential amino acid, meaning that the body may not be able to synthesize all it needs when it is under physical stress.

Effects on GH

Glutamine is the latest amino acid to generate excitement as a GH-releaser thanks to a 1995 study by Thomas C. Welbourne of Louisiana State University College of Medicine in Shreveport. Welbourne showed that a surprisingly small oral dose of about 2 grams of glutamine raised growth hormone levels more than four times over that of a placebo. Even more exciting, age did not diminish the response at least in this small study of volunteers, who ranged from thirty-two to sixty-four years. The only person who did not respond was a thirty-two-year-old obese woman. "Although this represents a small increment in circulating growth hormone," says Welbourne, "note that it is effective in eliciting growth hormone's metabolic effects." The dosage of 2 grams, he says, represents a "window" of effectiveness, since a smaller dose would probably not be enough to raise the levels of circulating glutamine needed to stimulate growth hormone and a larger amount would accelerate the removal of glutamine from the blood by the liver.

Anti-aging benefits

Glutamine is the amino acid that is most used by the body, particularly during times of stress. The immune system and the gut practically live on glutamine. If the body does not produce enough glutamine, muscle loss and immune dysfunction can occur. The gut atrophies, meaning nutrients all kinds cannot be absorbed as well as before.

A 1993 study by Welbourne in animals showed that glutamine supplementation protects muscle mass and prevents acidosis, which occurs with strenuous exercise and causes muscle breakdown. According to Judy Shabert, M.D., author of *The Ultimate Nutrient Glutamine*, supplementation with glutamine, especially in times of stress, would prevent muscle wasting. In a foreword to the book, Douglas Wilmore, M.D., of Harvard Medical School, points out that glutamine is a key to the metabolism and maintenance of muscle, the primary energy source for the immune system, and essential for DNA synthesis, cell division, and cell growth, all factors that are enhanced by GH. It also crosses the blood-brain barrier into the brain, where it increases energy and mental alertness.

High levels of glutamine in the blood translates into greater health as a 1994 study showed. In a survey of thirty-three people over the age of sixty, those at the top of the scale of blood glutamine levels had fewer illnesses, lower cholesterol, lower blood pressure, and were closer to their ideal weights than people at the bottom of the scale in this nutrient. The low-glutamine subjects had higher rates of arthritis, diabetes, and heart disease, while those who were high in glutamine said that they felt great.

Clinical usage

An effective growth hormone release. Available in pills and an almost tasteless powder form in which one teaspoon equals 2.5 grams. According to Giampapa, 2 grams of glutamine are more potent than 1 gram amounts of arginine, ornithine, and lysine combined.

Dosage

2 grams at bedtime. (See "A Note on Stacking the Amino Acids" below.)

Side effects

At recommended dosage level, it has virtually no side effects and is low in toxicity.

OKG (L-ornithine Alpha-Ketoglutarate): French Tickler.

What it is

This is strictly speaking not an amino acid, but a French designer nutrient that consists of ornithine bound to two molecules of alpha ketoglutarate, which is a precursor to glutamine and has GH-stimulating effects. It boosts glutamine levels even more than glutamine itself.

Effects on GH

OKG in doses of 10 grams or more was more effective than either ornithine or alpha ketoglutarate in raising GH release several fold in a majority of normal subjects.

Anti-aging benefits

In France it is used to rebuild body tissue after surgery, trauma, burns, and other catabolic (body-wasting) conditions.

Dosage

10 grams or more taken with fluids. (For supplier, see appendix.)

Side effects

Well tolerated if taken with fluids. It has not been widely tested in this country as a GH-releaser.

A Note on Stacking the Amino Acids

There is reason to believe combining glutamine, arginine, ornithine, and lysine will have a synergistic effect. Giampapa, who is medical director of the Longevity Institute International in Montclair, New Jersey, has followed the beneficial changes that GH agonists have made in hundreds of age-related hormone-deficient individuals. Based on his research, he recommends starting with GH agonists in order of their effectiveness. His beginner's stack is 2 grams of arginine, 2 grams of ornithine, 1 gram of lysine, and 1 gram of glutamine. This will have a minimal to moderate effect; then work up the ladder to 2 grams each of glutamine and lysine as well as arginine and ornithine. "That's going to make a noticeable increase in your GH secretion," he says.

The glutamine-arginine-lysine stack raises insulin as well as growth hormone. Generally when GH levels are rising, insulin levels are falling (see Chapter 17 for a full explanation of the relationship between insulin and GH), but if you increase the levels of insulin somewhat at the point when growth hormone levels are up, you can get a very high anabolic effect. According to a British study on rabbits, the combined action of insulin (which increases the rate at which amino acids enter the muscle), and growth hormone (which stimulates protein synthesis), produces just the right set of circumstances for incorporating amino acids into protein and thus building muscle. By taking them together at bedtime, you stop the body from breaking down as much muscle tissue as it would ordinarily do during the seven- to eight-hour period that you're sleeping.

Di Pasquale advises taking a glutamine, arginine, lysine amino acid stack at 3 grams each at bedtime and building a gram a week until you reach 5 grams each. This combination will raise growth hormone levels by about 10 to 20 percent based on extrapolations from animal studies, he says. The 15 grams of amino acids that you get from the supplements are about the equivalent of the protein in two eggs and just as safe for the body.

Glycine: Exercise Enhancer

What it is
A nonessential amino acid.

Effects on GH
Two studies found that this amino acid increased GH in the serum. In one, 6.75 grams at bedtime caused an three-fold increase, while a Japanese research team showed that 30 grams raised GH levels ten times over baseline in patients who had gastric surgery. An oral dose of 250 milligrams in normal volunteers also showed a significant, but less pronounced, rise in GH. They conclude that "the facts demonstrated that glycine is one of the stimulatory agents inducing the pituitary gland to secrete HGH." Glycine has also been found useful in increasing output in exercise workouts.

Anti-aging benefits
May be useful in dampening hyperactive brain activity that produces spasms. In one study, 1 gram of glycine a day for six months to one year significantly reduced spasms in all ten patients with severe chronic spasticity in the legs, including seven with multiple sclerosis.

Clinical usage
For some unknown reason, this amino acid has been completely overlooked as a GH-releaser and has not been researched as well as some of the others. Glycine is inexpensive, well-tolerated, and tastes sweet.

Dosage
250 milligrams to 6.75 grams.

Side effects
No reported toxicity or adverse side effects at recommended dosages. Taking glycine with trytophan is ineffective, since both nutrients compete for space on the same neuron receptor sites.

Tryptophan: A Real Sleeper

What it is

An essential amino acid that is a precursor to serotonin—the neurotransmitter that helps bring on sleep.

Effects on GH

Five studies all reported small increases in growth hormone with doses of 5 grams plus. In one study, an intravenous infusion of 5 to 7 grams over the course of fifteen minutes caused a marked elevation of serum GH between thirty and ninety minutes after administration. In the studies, only about half the subjects responded and all became drowsy.

How it works

Tryptophan is converted to serotonin, which increases growth hormone during sleep. The nutrient is also found in milk, which is why when your mother told you to drink hot milk before retiring at night, she knew what she was talking about.

Anti-aging benefits

Useful for inducing sleep and reducing effects of jet lag. It appears to be a mind-body regulator, decreasing anxiety and depression and with the addition of vitamin B6 may reduce severity of panic attacks.

Clinical usage

Useful only at bedtime since it makes you sleepy. For best results, it should be taken with B6 (30 milligrams) and vitamin C (250 milligrams), which the brain uses to convert tryptophan to serotonin.

Dosage

500 milligrams to 2 grams at bedtime.

Side effects

Drowsiness, headache, sinus congestion, constipation. Should not be used before driving.

Cautions

The FDA withdrew tryptophan from the market as a dietary supplement when it was linked to a rare blood disorder, eosinophilia myalgia, in some users. These cases were traced to some impurities in a few batches made by a Japanese company and there have been no reported cases since then. Although a rare reaction to tryptophan cannot be completely ruled out, it appears almost certain that the problem was due to contamination. Tryptophan is available only on prescription and can also be ordered directly from compounding pharmacies (see Appendix).

OTHER NATURAL SUBSTANCES

Niacin: A Hearty Vitamin

What it is

Niacin is commonly used to refer to the two forms of vitamin B-3, nicotinic acid and nicotinamide.

Effects on GH

Niacin is a potent GH-releaser. Two scientific studies show that 200 milligrams of niacin given intravenously increased GH levels eight-fold with the GH peak occurring two hours after the administration. It did not stimulate GH in people who were one and a half to twice their ideal weight, which is not surprising since obesity blocks GH release (see Chapter 17). Pearson and Shaw state that a 200-milligram dose of niacin "will cause a mild GH release—about the same as a nonexercising teenager."

How it works

Niacin is believed to play a role in the release of GH, particularly in its most potent form called xanthinol nicotinate, which passes more easily through the membrane of cells and through the blood-brain barrier.

Anti-aging benefits

This is an anti-death vitamin that significantly reduced mortality in every cause of death analyzed by a milestone study in the seventies and eighties called the Coronary Drug Project, including cardiovascular disease, cancer, and all "other" causes. Nicotinic acid at doses of up to 7 to 8 grams outperformed several of the major cholesterol-lowering drugs, which were shown to be no better than a placebo in extending life. Other studies have demonstrated that niacin actually reverses atherosclerosis and lowers serum cholesterol by an average of 25 percent and triglycerides by up to 52 percent and raises HDL levels by 33 percent. And it is also a cognitive enhancer, improving memory by 10 to 40 percent in both young and middle-aged normal, healthy subjects as compared with controls in a double-blind study.

Clinical usage

Niacin is effective on its own or as part of an amino acid stack. (See "A Note on Stacking the Amino Acids" above.) "With just using L-glutamine and niacin, you can get a 30 to 40 percent elevation in IGF–1 levels," says Giampapa.

Dosage

200 milligrams to 1 gram (see note below).

Side effects

Niacin can cause a harmless, though initially scary, reaction known as flushing, when the skin turns red, feels very hot, and may itch or tingle. This can happen in doses as low as 100 milligrams taken on an empty stomach and is caused by dilation of the arteries and a release of histamine. The reaction is harmless, generally subsiding within twenty minutes.

Note

To minimize flushing, start with small doses and slowly build to 1 gram. Tolerance to flushing usually develops within one week. If niacin is used alone as a GH-releaser, taking it with food will also minimize flushing. If using it in a stack with arginine, take it on an empty stomach at bedtime. You may be asleep by the time the flushing occurs. If not, lie back and enjoy it! The niacin flush is basically the same as the "sex flush" documented by Masters and Johnson that occurs with histamine release during orgasm.

Caution

Doses over 800 milligrams with timed-release forms of have been linked to liver damage.

GHB (Gamma Hydroxybutyrate): The Sexy Nutrient

What it is

A naturally occurring substance found in every cell of the human body as well as the brain. It is both a precursor and a breakdown product of GABA (gamma-aminobutyric acid), a neurotransmitter that is involved in the regulation of the anterior pituitary gland, which secretes growth hormone.

Effects on GH

GHB is one of the most potent stimulators of growth hormone from the pituitary that has been discovered. A Japanese study of GHB found that a dose of 2.5 grams given intravenously to six young healthy men increased the levels of growth hormone ninefold within thirty minutes and rose to sixteen times over the baseline within one hour! In the same study, GHB also increased by fivefold the hormone prolactin, which is involved in lactation. Generally a rise in prolactin is considered undesirable since it works in opposing ways to growth hormone, for example increasing fat and lowering immunity and sex drive. Many of the physiological factors that induce growth hormone, such as sleep, stress, hypoglycemia, and exercise, also increase the secretion of prolactin. But the rise in GH was so much greater that it probably overwhelmed any negative effect that the rise in prolactin might have had. The stimulation of growth hormone no doubt accounts for its reputation among bodybuilders as one of the best fat-burning, muscle-building compounds.

How it works

GHB can double the concentrations of the neurotransmitter dopamine in the brain. The release of growth hormone is under dopaminergic control in the hypothalamus and drugs that induce dopamine in the brain, such as L-dopa and bromocriptine (see below), are potent stimulators of GH. Unlike those drugs that work in a straightforward manner, GHB appears to take a circuitous route by *inhibiting* the release of dopamine from the nerve endings. This has the effect of creating higher levels of dopamine within the brain cells where it is stored. Some scientists have speculated that even after GHB wears off, the excess dopamine can be released into the synapses between nerve cells, leading to greater dopamine activity the next day.

Anti-aging benefits

Although the evidence is anecdotal, GHB users testify to its potency as an aphrodisiac. Dr. Ahlschier, who has used GHB on himself, says: "It is very effective stimulator of growth hormone, a great sedative, and a true aphrodisiac," adding that "the most incredible sex is attained in the twilight zone of GHB." It also induces an extremely deep sleep at higher doses, which, like its effects on the libido, could be related to growth hormone release.

Clinical usage

GHB can be used safely at low doses to stimulate growth hormone (see side effects and cautions below). Or, according to Giampapa, 500 to 750 milligrams of GHB can be added to the stack of amino acids above, along with 1 to 3 milligrams of melatonin (see Chapter 15), both of which will aid in sleeping.

Dosage

500 milligrams to 1 gram at bedtime as directed by physician.

Side effects

At high doses, it can induce a state of sleep so deep that the person can not be easily aroused. It has also been abused by bodybuilders who have taken many spoonfuls of the compound, with side effects, such as muscle spasms and twitching, associated with overuse. This led to the FDA withdrawing it from the market in 1990, although it is now generally available in health food stores.

Cautions

Buy pharmaceutical grade version manufactured by a known laboratory. (See suppliers in Appendix.)

PRESCRIPTION DRUGS

For the most part, we recommend that you use nutritional supplements found in nature rather than synthetic drugs to stimulate growth hormone release. But there are a number of prescription drugs that can enhance the growth hormone response. These are drugs that thousands of people have used for many years for anti-aging purposes without any serious side effects. While these are all FDA-approved drugs whose safety and reliability has been confirmed in hundreds of thousands of patients, none of them has been approved for the purpose of raising GH levels. And because prescription drugs are usually synthetic and not found in nature like the nutrients listed above, they tend to have more side effects and more serious ones. But we are including them here for several reasons. First, some of them are quite effective in raising GH levels. Second, you may have read about their use elsewhere. Last, but not least, we believe that you should know what all your options are and make up your own mind based on the best available information. The goal is to stimulate replacement levels, not superphysiologic levels, of growth hormone. Remember, none of this should be tried without close monitoring by a physician specializing in anti-aging medicine. It is imperative that you read carefully the dosage levels, side effects, cautions, and risks associated with their use.

FOUR GH-RELEASING ANTI-AGING DRUGS

The following four drugs have long been used worldwide for life extension purposes. Recent studies have shown that all four significantly raise growth hormone levels in people, and in some cases restore the release of GH in elderly people back to youthful values. At the low dosages in which they are used for GH stimulation and anti-aging, no serious side effects are associated with their use. And while there are risks involved in chronic use—even chronic use of low-dose aspirin can be injurious to health—the benefits in terms of GH release appear to outweigh the risks.

L-dopa or Levadopa (Sinemet: The Longevity Awakener)

What it is

This is actually an amino acid that occurs naturally in the human body. It is available only by prescription for use in Parkinson's disease.

Effects on GH

L-dopa is one of the most effective GH-stimulators in both animals and humans. A 1976 National Institute on Aging study by Joseph Meites, one of the most distinguished researchers in hormones and aging, found that .5 gram of the drug per day increased the GH output of men over sixty who did not have Parkinson's to levels approaching that of young adults! And there were no adverse effects at this dosage. A 1982 study by Sonntag and his colleagues found that L-dopa restored the growth hormone pulses that had decreased in old rats to those of young rats. According to Pearson and Shaw, .25 to .5 gram of L-dopa can cause GH release as shown by fat loss and muscle gain.

How it works

L-dopa is a precursor of dopamine, one of the important neurotransmitters in regulation of growth hormone. Drugs that raise the levels of dopamine in the brain are among the most powerful stimulants of GH release. It is also a building block for the transmitters norepinephrine and epinephrine, which have all been shown to increase GH secretion.

Anti-aging benefits

L-dopa attained cinematic fame in *Awakenings* as the drug that jolted patients out of their fifty-year trance caused by a form of sleeping sickness. It may also serve to awaken—although not as dramatically—older people from the lethargy of aging. A number of studies in animals reveal that L-dopa has remarkable anti-aging and life-extending effects. In a classic longevity study, Dr. George Cotzias, the developer of L-dopa for Parkinson's disease, fed three groups of mice varying amounts of L-dopa—one milligram, 20 milligrams, and 40 milligrams, respectively, per gram of diet. A control group had no L-dopa in their diet. The group on the highest dosage—40 milligrams—lived the longest, with almost twice the number of animals in this group still alive at 18 months compared with the untreated controls (73 percent versus 39 percent). Although the 40-milligram group had the highest initial mortality, perhaps because they were hit with too high a dose too fast, they also lived the longest in terms of maximum life span—the benchmark of longevity studies (more than 1,000 days compared with 950 for the controls).

Of course, the question is how well the old treated animals function, since an extended old age spent hobbling around with a cane is no bargain. L-dopa turned old rats into young athletes in one study. While rats are known for their ability to leave a sinking ship, old rats don't do so well, tiring easily and sinking in tests of swimming ability. But the L-dopa–fed rats had the same level of stamina, efficient swimming posture, and use of body movement as the untreated young controls. Dopamine controls motor control and coordination, which may account for why the old animals regained their youthful swimming, but at least some of its restorative abilities may also be due to increased secretion of growth hormone.

In his book *The Neuroendocrine Theory of Aging and Degenerative Disease*, Vladimir Dilman points out that L-dopa helps restore the sensitivity of hypothalamus to feedback from the hormonal signals from the rest of the body, which is impaired with age. This in turn improves the homeostasis of the body (see Chapter 5 for an explanation of Dilman's hypothesis). Since lack of responsiveness to hormonal signals could be involved in the decreased secretion of GH with age, the use of L-dopa may reverse this communication breakdown.

Some people who have used L-dopa as a GH-releaser and for life extension purposes claim that it is a great sexual stimulant— another well-known effect of GH. But, according to John Morgenthaler and Dan Joy in *Better Sex Through Chemistry*, the research shows that it tends to work best among those whose sex drive is low to begin with. In one study of male Parkinsonian patients, about half showed increased libido with L-dopa treatment, and the authors concluded that it improved sexual function provided that the endocrine system was still working.

Clinical usage

In low doses, far under those used for Parkinson's patients, L-dopa is effective in raising GH levels without side effects. Since it works through a different mechanism from arginine and ornithine, it may be a useful addition to amino acid stacks.

Dosage

125 to 500 milligrams at bedtime as directed by a physician; best taken as carbidopa (sinemet).

Side effects

This is a serious prescription drug that *must* be used with medical supervision. Side effects mostly seen at doses higher than 500 milligrams include dizziness, nausea, vomiting, involuntary body movements, orthostatic hypotension, high blood pressure, arrhythmia, confusion, psychosis, depression, and gastrointestinal bleeding.

Cautions

Greater adverse effects with pure forms (not carbidopa) or at high doses.

Hydergine (Ergoloid Mesylates): Brain Rejuvenator

What it is

An ergot derivative closely related to bromocriptine, hydergine is prescribed for age-related decline in mental capacity, Alzheimer's, and other dementing conditions.

Effects on GH

Hydergine is one of the few GH-releasers that has actually been shown in clinical testing to be active in elderly people. In a 1983 study by Ermanno Rolandi and his associates at the University of Genoa in Italy, long-term treatment with hydergine caused a significant increase in the nocturnal peaks of growth hormone of five men and five women between the ages of seventy-two and seventy-eight. These were patients who were hospitalized for chronic cerebral vascular insufficiency—an indication for the use of hydergine—but were otherwise in good condition. After one month of therapy with hydergine at a dose of 6 milligrams per day, the nocturnal peaks of growth hormone increased to almost twice the levels they were before treatment. At the same time it sharply lowered prolactin, which is generally considered anti-aging, but did not raise the levels of the stress hormone, cortisol.

How it works

It is believed to work through its stimulation of dopamine, but it also affects noradrenalin, another neurotransmitter that stimulates GH. The drug is actually a mixture of ergot derivatives, one of which, dihydroergocornine, is one of the strongest GH-releasers of the various ergots.

Anti-aging benefits

Hydergine gained wide popularity after Pearson and Shaw touted it in their book as a "smart" drug. According to them, it helps prevent or correct aging in the brain by increasing protein synthesis in the brain, which is required for memory; stabilizing brain EEG energies under conditions of low oxygen supply; slowing the rate at which the age pigment lipofuscin accumulates in the brain; improving memory and learning; and stimulating the growth of neurites (nerve-cell connections required for forming new memories, which are lost with age). It has reversed at least in part some brain damage resulting from stroke, infections, radiation, and some birth defects. In one double-blind clinical study, 12 milligrams per day for two weeks increased intelligence. It has also been effective in treating bronchial asthma and tardive dyskinesia, the Parkinsonian-like symptoms that are a side effect of long-term use of antipsychotic drugs. People who have used it for anti-aging purposes report that it increased feelings of well-being, energy levels, and fat loss. Considering that almost all these changes are similar to the ones seen with GH replacement, it is likely that many of the effects attributed to hydergine are due, at least in part, to the restoration of youthful growth hormone levels in later life.

Clinical usage

This is a "best bet" for prescription drugs used to stimulate growth hormone. It is extremely safe and has been used on a daily basis for decades by older people in Europe to preserve brain function and increase energy levels. Hydergine, like L-dopa, seems to have significant effects for several months, although there are no long-term studies on people to see if the effect is sustained for a longer period of time.

Dosage
One milligram 3 times a day for memory enhancement. Six milligrams found effective for GH release. Can increase up to 12 milligrams as directed by a physician.

Side effects
Rare, but may include nausea, drowsiness, slow heart rate, rash.

Cautions
Can potentiate the effect of caffeine, causing headache and insomnia. Build up dosage level slowly.

Clonidine (Catapres): Antihypertensive Agonist

What it is
A blood pressure–lowering drug that is also used for many diverse conditions, including Tourette's syndrome, migraines, ulcerative colitis, painful or difficult menstruation, and hot flashes from menopause, and to promote growth in children whose growth is delayed.

Effects on GH
In animal and human studies of the acute response to clonidine, it effectively stimulates GH secretion. There have been conflicting results with clonidine in older animals. In one study, older rats had a reduced GH response to clonidine compared with younger animals. But researchers at the University of Milan found that the ability to respond to clonidine is not lost with age. When the Italian scientists gave injections of clonidine to old beagle dogs along with growth hormone–releasing hormone (GHRH), the GH peaks were more frequent and higher than with GHRH alone. And this response was only slightly reduced in old dogs compared with young adult ones.

It is still not clear whether clonidine works to raise GH levels over the long term. Two clinical studies found that hypertensive patients on clonidine treatment had basal GH levels that were no different from those of untreated controls who did not have high blood pressure. Although in one these studies fifteen minutes of treadmill exercise raised the GH levels by four to five times over baseline, the clonidine-treated patients did no better than the controls. However, Stephen Borst and his associates at the Department of Veterans Affairs Medical Center in Gainesville, Florida, who reviewed these studies, point out that the research made only single measurements of growth hormone and did not examine the GH pulses. As we have mentioned, it is the frequency and amplitude of the GH pulses that are important in producing the anti-aging effects.

The Florida scientists also note that in animal studies chronic use of clonidine raised pulsatile GH levels. Two groups reported increased pulsatile GH secretion in rats. And the University of Milan scientists found that in old dogs, fourteen days of clonidine administration increased the frequency and amplitude of the spontaneous growth hormone bursts as well as the overall GH secretion. In fact, the secretory patterns of growth hormone, say the researchers, was indistinguishable from that of untreated young dogs. In other words, clonidine completely *reversed* the loss of GH release that occurs with age!

Anti-aging benefits
It appears to have some potential as a "smart" drug. In one study it improved the mental function of some alcoholics with Korsakoff's psychosis, alcohol-induced brain damage; patients had better recall, were more alert, and had improved learning ability.

Clinical usage
Clonidine works to lower blood pressure by another route, stimulating a set of nerve endings called alpha-adrenergic blockers. While its effectiveness in raising GH has not yet been established in clinical studies, the Milan studies in animals show that it may have the potential for reversing the loss of GH pulsatility with age.

Dosage
0.1 to 2.4 milligrams as directed by physician.

Side effects

In dosages that are in the clinical range, it is safe and well tolerated. Most common side effects are dry mouth, drowsiness, and sedation. Other common side effects include constipation, dizziness, headache, and fatigue. Less frequently reported are orthostatic hypotension (low blood pressure upon standing), loss of appetite, nausea, vomiting, weight gain, pain or swelling of glands in the throat, breast pain or swelling, arrhythmia, rapid heart rate, EKG changes, painful blood vessel spasm, diminished libido, elevated blood sugar, and general feelings of ill health.

Cautions

It can not be taken with a beta blocker as the combination has caused blood pressure to *rise* in some people. Abrupt withdrawal from clonidine may cause rebound high blood pressure.

Dilantin (Phenytoin): The "Dreyfus" Drug

What it is

An anticonvulsant drug, related to barbiturates, used to prevent seizures in epileptics.

Effects on GH

The studies on Dilantin show varied results from no change to a significant elevation of growth hormone. One 1980 Finnish study found that women epileptic patients treated with phenytoin as well as another anticonvulsant, carbamazepine, had serum GH levels that were more than twice those of the controls, but for men there was no difference. A 1984 Italian study showed that Dilantin probably affects the hypothalamus-pituitary regulation of growth hormone. In this study five epileptic patients had a higher growth hormone response to L-dopa after one month of treatment with Dilantin.

In the most interesting study from the point of view of using Dilantin as a GH-releaser, a group of New Zealand researchers gave six healthy subjects 500 milligrams of the drug at night and then had them exercise on a stationary bicycle at 100 watts for forty minutes the following morning. The researchers also measured the GH levels of the same subjects doing the same exercise at the same intensity without Dilantin. Both times, their growth hormone concentrations rose at the end of the exercise, but after Dilantin, the peak was considerably higher. The study also found that Dilantin plus exercise caused an increase in free fatty acids and glycerol, which suggests, say the investigators, that Dilantin increases lipolysis, or the destruction of fat cells.

Anti-aging benefits

In his two books, *The Lion of Wall Street* and *A Remarkable Medicine Has Been Overlooked*, Jack Dreyfus recounts the story of how Dilantin stopped his chronic depression dead in its tracks. As financial wizard Dreyfus points out in his books, Dilantin has an amazing range of clinical uses from treating anxiety and depression to treating heart disease, neuromuscular disorder, pain, ulcers, asthma, alcohol and tobacco addiction, and a host of other conditions. It is also a "smart" drug that has been shown in laboratory animals to increase learning ability and extend life span. Anecdotally, people have reported using Dilantin at low dosages of 50 milligrams per day to increase mental quickness, memory, and ability to concentrate with no adverse side effects.

Dreyfus began taking Dilantin soon after he became depressed in 1958. Recently I shared a cheese sandwich with the "Lion of Wall Street" in his Madison Avenue office. A tall, sprightly eighty-three-year-old with a commanding presence, Dreyfus is bright, alert, and involved. He has dedicated his life to championing the multiple benefits of Dilantin and was extremely interested in what I had to say about Dilantin as a growth hormone releaser. Since growth hormone has both an anti-anxiety and calming effect along with enhancing memory and cognition, it could be that at least some of the positive effects on brain neurotransmitters attributed to Dilantin are GH-related.

Clinical usage

Side effects do not appear to be a problem in the very low dosages used for increasing intelligence, altering mood, or anti-aging purposes. It may be particularly helpful as a GH-releaser before exercising.

Dosage

50 to 400 milligrams a half hour to one hour before exercise as directed by physician.

Side effects

Although Dilantin appears to be safe at low dosages, it is still a drug and should be used with caution. The drug is usually well tolerated, and the Dreyfus Medical Reasearch Foundation reports that the few side effects associated with this drug are reversed when the drug is stopped. The following side effects have been reported in the *Physician's Desk Reference*: unusual growth of gums, nystagmus (uncontrolled eye movement), uncontrolled twitching, double vision, headaches, dizziness, insomnia, nervousness, confusion, slurred speech, fatigue, depression, nausea, vomiting, diarrhea, constipation, fever, rashes, hair loss, weight gain, joint pain, elevated blood sugar, coarsening of facial features, lip enlargement, hirsutism, Peyronie's disease, liver damage, blood disorders.

GROWTH HORMONE–RELEASING PRODUCTS

These are nutritional products that have been specifically formulated to increase growth hormone release. In effect, they have done the work of stacking the nutrients for you by putting together dosages that their own laboratories have shown work best together. According to the manufacturers, they have no side effects associated with their use. All the nutrients contained in these products have already been reviewed in this section, with the exception of the patented GHRP (growth hormone–releasing peptide) in AminoTropin-6.

Natural Growth Hormone Releasing Formula

The Life Extension Foundation offers most of the growth hormone–releasing nutrients discussed in this book, including arginine, ornithine, lysine, glycine, and glutamine. The foundation also offers several growth hormone–releasing multinutrient formulas, including Daytime Growth Hormone Releasing Formula, Nighttime Growth Hormone Releasing Formula, and Growth Hormone Releasing Powder Formula. The most comprehensive of these formulas is called Natural Growth Hormone Releasing Formula, which contains arginine, ornithine, lysine, alpha-ketoglutarate, glutamine, and xanthinol nicotinate. (See Appendix for supplier.)

HGH Plus—PM

This new formulation made by the Weider Nutrition Group combines eleven nutrients including amino acids into a high-powered nutritional alternative to growth hormone injections. This supernutrient is a combination of L-glutamine, ferulic acid, chromium, OKG, L-arginine pyroglutamate, L-lysine, arginine aspartate, vitamin B6, pantothenate, L-tyrosine, and xanthinol nicotinate.

Two of the top researchers in nutritional ergogenic aids for sports, Luke Bucci, Ph.D., and Jerzy Meduski, Ph.D., designed the formula based on the latest scientific information and their clinical experience. They have combined these nutrients into a single packet so the right amount of each one use matches the specific amount used in the research studies. So, for instance, Isodori in the study cited above used 1.2 grams of L-arginine pyroglutamate and 1.2 grams of L-lysine to achieve a synergistic effect. Isodori also found that the amino acid combination increased IGF–1 levels, proving that the GH was physiologically active. Other combinations of ferulate, tyrosine, ornithine, arginine aspartate, and OKG have increased growth hormone stores so that they can be released normally when the body needs GH the most, during sleep or after exercise. Modest amounts of arginine and ornithine have increased lean body mass and strength in middle-aged men, presumably because of the release of GH. According to the formula's developers, Bucci and Meduski, "There is substantial evidence that this combination of amino acids and other key nutrients can actually make growth hormone more effective in the body." (See Appendix for supplier.)

AminoTropin-6

This new formula, made by Gero Vita International, is the first over-the-counter food supplement designed to combine nutrients with a patented GHRP (growth hormone–releasing peptide). The GHRPs, which were first developed in this country by Dr. Cyril Bowers of Tulane University in New Orleans, are the leading edge of oral drugs that are proving to be as powerful as injections of HGH. In addition to the GHRP, AminoTropin-6 contains GABA, arginine, lysine, and xanthinol nicotinate. According to Gero Vita scientists, the product can boost levels of GH by up to 50 percent. And, they say, because the compound induces the pituitary to release GH naturally, there are no adverse side effects. In fact, they contend it is even safer than taking GH itself. For much more on the development of AminoTropin-6 and other GHRPs, see Chapter 21. (See Appendix for supplier.)

THE IMPORTANCE OF CYCLING

There are two reasons that you should cycle GH-releasers, that is, use them for a period of time and then go off them for a period of time. First, all hormones in the body work via delicate, complex feedback mechanisms, with one hormone signaling another to turn on or off in response to the amount of hormone circulating in the body (see Chapter 4). By continually supplying the hormone or anything that stimulates the hormone from the outside, you can interrupt the natural feedback mechanism. This goes for injections of HGH (see Chapter 19) as well as growth hormone releasers. Second, the GH-releasers cause increased activity in the receptors that bind to them. Stopping the releasers for a period of time gives the receptors a chance to rest.

"The state of our knowledge right now in terms of replacing or stimulating growth hormone," says Giampapa, "is that we know that the short-term effects are very positive, but we don't know about the long-term use with problems of inhibiting the normal feedback mechanism to the brain." There are a number of ways to cycle, which we give below, but the concept is the same. In this book, we offer you the best of what anti-aging research has to offer. But the optimal use of GH-releasers is a work in progress.

TWO VIEWS HOW TO CYCLE

"At the Longevity Institute International," says Giampapa, "we make a blanket recommendation that anything you use to increase growth hormone should be done for three weeks and then stop for a week. In that way you don't completely shut off your feedback mechanism while you're taking the substance. Then you get a whole week off for your body to make sure it resets itself. This is what we recommend until we have a better idea of what the feedback mechanisms are and how they actually work."

According to Di Pasquale, "there is an art and science to cycling." Except for basic vitamin, mineral, and antioxidant supplements, he advises cycling almost any substance. "Your body will adjust to most things and the response you're looking for attenuates. By cycling, you start getting the response afresh." If you are using nutritional compounds, such as an amino acid stack, he recommends taking it every day for four to six weeks and then cycling off for two weeks. With drugs like clonidine that have a long-lasting effect, he advises going off a longer period of four to six weeks.

If you are doing exercise training as well as taking growth hormone stimulants—a combination that will give you the best effect—then that should play a role in how you cycle, he says. "I try to make a synergistic effect between a person's life style, exercise training, nutrition, and diet. For example, I tell people to cycle their nutritional supplements according to the training they do. So if they are starting a 12 week program, they start off light and don't take any supplements for the first two weeks. On the third week, they start taking the supplements, raising the amounts in the fourth and fifth weeks, and then take the full amount until the end of the program. Then they go off it for a few weeks until the next cycle. The idea is to take more when you're training harder so you get a synergistic effect and then drop back when you're not working out so hard. In this way you maximize a person's usual fitness against their muscle mass."

ANTI-AGING OVER THE LONG HAUL

Think of growth hormone agonists as only the first step in a road that winds into a brighter and brighter future. There are already hormones and drugs that are now being tested in the clinics that will have an impact far beyond anything that is described in this chapter. Many of these should be available to the general public within a few years or even sooner in the case of AminoTropin-6. Beyond that, there are even more exciting substances that will propel us into a new age of agelessness and longevity unlike anything we have ever known.

In the next two chapters, we will show you how to get the optimal benefit from GH-releasers (or injections of HGH) by combining them with a program of diet and exercise.

Maximizing the Effect: Diet and Nutrition

Raising your growth hormone and IGF–1 levels will melt fat and build muscle *without diet or exercise*. This is true provided the levels of your own hormone have fallen below what they were in your twenties or thirties—which means most of us over the age of thirty-five or forty. In the Rudman study, the twelve old men did not change their eating or exercise habits, yet they lost 13 percent fat and increased their lean body mass by 8.8 percent. A 1996 double-blind study of elderly men with GH replacement carried out by Maxine Papadakis and her associates at the University of California in San Francisco found similar favorable changes in body composition of a 13 percent decrease in fat and a 4 percent increase in LBM, again without cutting down on calories, doing aerobics, or pumping iron.

But our Growth Hormone Enhancement Program, combining GH agonists, hormonal dietary manipulation, and GH-releasing exercise, will increase and accelerate these benefits dramatically. You can become a shape-shifter, tapering your waist, slicing fat off your belly, smoothing your thighs and buttocks, toning and strengthening your muscles so that you look better at forty, fifty, sixty, seventy, and beyond than you looked at nineteen. And you will gain all the other anti-aging benefits as well that are documented in this book.

In this chapter, we will show you how to lose fat and gain lean mass without counting calories or giving up the foods you love. And we'll show you how to do it safely and easily. You can even use this diet to gain fat-free weight as your muscle mass increases. In the following chapter we will show you how exercise can be used to maximize the growth hormone response for the greatest possible age-reversing effects.

A WEIGHTY MATTER

The biggest reversible obstacle to growth hormone release is body fat. In fact, obesity is as great a factor in the decrease of growth hormone as aging. A 1991 study by Dr. A. Iranmanesh and his colleagues estimated that growth hormone fell 6 percent for each unit increase in body mass index (BMI—a measurement of height to weight, which is equal to the weight in kilograms divided by the square of the height in meters. See BMI chart in Chapter 13.) According to the researchers, two factors reduced the rate of twenty-four-hour production of growth hormone—going from age twenty-one to forty-five years or increasing the body mass index from 21 to 28. Or to put it another way, the lower the percentage of body fat, the higher the level of growth hormone.

The more body fat you have, the lower the peaks of the secretory bursts. Pointing to the literally ballooning rate of obesity in the United States, he says, "If your pituitary gland is intact, you can increase your level of growth hormone without taking any drug whatsoever. And that is with exercise and decreased caloric intake. Age is not the only bad guy; it is also body fat. Starvation, exercise and decrease in body fat are the three things that start your own GH secretion going."

WHAT GOES AROUND COMES AROUND

Here is another great incentive to lose weight, added to all the ones you've always known about, such as looking great; lowering your risk factors for heart disease, hypertension, diabetes, and cancer; and extending your life. As you lose fat, you get an augmentation of growth hormone levels that, in turn, helps you shed more fat. And if you increase your muscle mass through resistance training (see

Chapter 18), you also release high concentrations of growth hormone, which help you build more muscle.

THE FEAST-FAST CYCLE

The best explanation for how growth hormone works to reduce weight and build muscle, according to Danish researcher Jens Sandahl Christiansen, comes from a paper on the metabolic action of growth hormone that was published in *Nature* in 1963. "Not many people have paid attention to it," says Christiansen, "but it explains a lot of the physiological things that we have encountered when we look at our patients."

The way it works is this. With every meal, we go through a three-stage cycle of feast and fast. In the first hour after eating (stage one), the blood sugar rises and insulin is released, which encourages the storage of excess carbohydrates and fat. After the second hour (stage two), growth hormone is released and the levels of insulin and blood sugar start to fall. At this stage, growth hormone acts to build up muscle protein, an activity that is enhanced by the presence of

Variations in Insulin and HGH During Feast-Famine Cycle

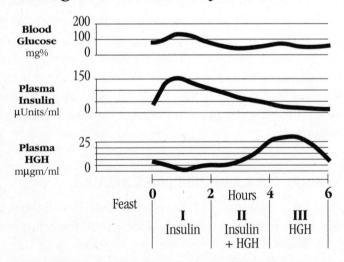

Suggested variations in plasma insulin and HGH concentrations during one feast-famine cycle.
Modified from: Rabinowitz, D., and Ziebler, L.K., *Nature* 199 (1963): 913–15.

insulin. In stage three, known as the postabsorptive phase, which occurs more than four hours after eating (the fasting stage), the growth hormone concentrations remain at a high level, while the insulin almost disappears. At this stage, growth hormone acts solely to mobilize the body's fat stores for burning as fuel.

RESTORING THE GH-INSULIN BALANCE

When we're young, our levels of growth hormone are high in relation to insulin. This is good because insulin, as we saw in Chapter 8, works to creates fat (lipogenesis), while growth hormone works to break down fat for use as energy (lipolysis). Growth hormone acts as a brake on insulin, keeping its fat creation and storage at a minimum. This is why, when you were young, you could pig out on pizza and fries and shakes all you wanted without paying the consequences.

Now, let's look at what happens when you age. Starting around your mid- to late thirties, even in the twenties for some people, you start to put on fat even though your diet and physical activity haven't changed. The reason is that the balance between growth hormone and insulin has shifted. Growth hormone has declined while the levels of insulin, if anything, have remained the same or even gone up. Insulin is an essential hormone required for the metabolism of sugar. Without it, we die. Insulin also stimulates muscle growth but to a much lesser extent than GH. But with less growth hormone around, insulin is free to turn every calorie you don't immediately expend into fat for later use. Insulin and growth hormone are both our friends; it is the balance between them that is the problem. By bringing growth hormone back to youthful levels—not beyond, which would have negative effects—we can restore the balance between GH and insulin so that they both work together as outlined in the feast-fast cycle above. In this way, when GH and insulin are both in the bloodstream together, they help to stimulate protein synthesis and muscle mass growth. At the same time the higher levels of GH work to block insulin's fat storage effect. And during the postabsorptive phase when insulin disappears from the bloodstream, growth hormone can melt away fat without interference.

GH AND FASTING

The history of fasting goes back to prehistoric times when humankind was dependent on an unstable food supply. The bodies of all animals are designed to function well during periods of short-term famine—the condition that obtained on this planet for human beings until the advent of agriculture about 10,000 years ago. Fasting has been used for centuries to maintain health, prevent digestive problems, and promote longevity.

The idea behind fasting for health purposes is a simple one. It allows the inside of the body to take a bath, cleansing itself of all its toxins. The body can store only a limited quantity of foreign matter. In addition to foods that require digestion, we take in the detritus of civilization in the form of artificial colors, preservatives, flavorings, pesticides, etc. These place a strain on the kidneys, bowels, lungs, liver, skin, and nervous system. Call it the Revenge of the Twinkies.

By ceasing to consume food, we give our digestive system a rest and eliminate the accumulated foreign substances from our body. But fasting has yet another function that has not been appreciated up to now. It is one of the most potent inducers of growth hormone. If the fast continues for several days, the concentration of GH rises to the excessive levels seen in people with acromegaly (giantism). It may be that the beneficial effects of controlled fasting, as well as the life-extension effects of food deprivation, are due at least in part to elevated growth hormone levels. (See Chapter 5). Later on in this chapter, we'll tell you how can make this feast-fast cycle work for you in our Growth Hormone Enhancement Diet.

GETTING STARTED

Walk into any bookstore as we have done and look at the shelves of diet books. There are stacks and stacks of them, but if you start thumbing your way through several of these books, you find a most curious thing. The vast majority of these books are much closer to religion than to science. Each one claims to have found the one true path to salvation and insists that all other paths are false. Yet the dietary principles are at complete variance with one another. One tells you to load up on carbohydrates and go easy on fats and protein, while the next one tells you that proteins and fats are the good guys and carbs are the villain. Other books push juicing as the way to health heaven while still other books hold that juice is the road to diet hell. And while many of the latest books are careful to include the latest research on metabolism and hormones, especially insulin, almost none of them mention the role of growth hormone. But as we have seen over and over in this book, *growth hormone is the key both to losing fat and to gaining muscle.*

The Growth Hormone Enhancement Program not only melts away fat, *it alters your body composition.* You may not change your weight, but you will change your clothes. Women may buy their first bikini in years as their waists and tummies shrink and their thighs become toned, their breasts get less droopy, their belly buttons rise (see Chapter 9). Men may take in their belts at the same time they let out their collars and increase their jacket size.

The Growth Hormone Enhancement Program can be used to lose pounds as well as fat. A majority of the American population is overweight, that is, more than 20 percent above their ideal weight. But remember that the ratio of fat-to-lean body mass is even more important. You can measure that using the self-tests provided in Chapter 13. If you are overweight for your build and you wish to lose pounds as well as body fat, the Growth Hormone Enhancement Program will help you achieve your goal much more rapidly and successfully than you have in the past. Just follow the principles outlined in this chapter and the next, while taking your growth hormone releasers.

For most people, this will be enough to bring about the desired body changes. But if you want to go further, especially if you are 50–100 percent or more over ideal weight, you will have to limit your caloric intake and increase your exercise level.

The way to do that is simple:

RX FOR WEIGHT LOSS

DECREASE CALORIC INTAKE BY 500 CALORIES PER DAY.

INCREASE PHYSICAL ACTIVITY BY 300 TO 500 CALORIES PER DAY.

If you follow this prescription, you will run up a deficit of 800 to 1,000 calories per day, or 5,600 to 7,000 calories per week. You can then expect to lose between 1.6 and 2 pounds of fat per week. You don't want to lose weight any faster than that since it will just trigger an evolutionary genetic strategy, called the starvation reflex, designed for periods when food was in short supply—that is, to hang on to every bit of fat in the cells. A complex interaction of nature and nurture, genes and eating habits, combine to create a set point, a kind of built-in equalizer that tries to maintain a stable body weight. Once the body has fixed on a set point, the metabolism does all it can to keep it at that point. If you go below the set point, the starvation reflex kicks in and the body lowers the metabolism to burn food more slowly and at the same time increases the appetite. As the weight goes up again toward the set point, the metabolism waxes and the appetite wanes. The Growth Hormone Enhancement Program outlined in this section represents the best way to reset that set point. (Soy extract is also helpful in changing the set point, see below.)

Burning 300 to 500 calories per day is equivalent to about three to five miles of running, one to two hours of vigorous weight training, or forty-five to sixty minutes of circuit weight training. See Chapter 18 for table of caloric expenditures of various common exercises.

THE FOUR FOOD GROUPS

When a prominent toxicologist was asked which foods he ate for his health, he said, "I eat them all. That way, I don't get too much of any one toxin."

Now *there's* a good reason for eating a wide variety of foods. Another is to get all the vital substances you need, listed in the next section. Eating from the four basic food groups is the easiest way to get it all. It is possible to be a vegetarian or a vegan (no dairy products or meats) and still be very healthy, but it requires a lot of careful planning to get all the protein, vitamins, and minerals you need. The four food groups are:

1. Fruits and vegetables
2. Grains and cereals
3. Dairy products
4. Meats

ELEMENTS OF NUTRITION

Once you have put everything into place—growth hormone releasers, exercise, and nutritional components, you may find that you're not eating less, but more! Remember that muscle weighs twice as much as fat, and as you lose fat and gain muscle mass, your body has to work harder to move that muscle mass around. In other words, you are expending more calories. And if you want to add to the muscle mass, you have to eat more.

First, let's take a look at the seven elements that make up all foods and are essential for any healthy diet.

THE THREE BASIC NUTRIENTS

Like ancient Gaul, all food can be divided into three parts—carbohydrates, fats, and proteins. The first two provide most of the fuel for the body to run on, while the third also provides the raw material for the cells and tissues.

Carbohydrates

- Carbohydrates have four calories per gram.
- The brain and nervous system run on the glucose provided by carbohydrates.
- Carbs fuel the muscles during exercise.
- Carbs help maintain tissue protein.
- The body converts protein to carbohydrates when carb stores are low.
- 100 to 150 grams of carbohydrate are needed to prevent drawing on the body's proteins.
- Carbs aid in the breakdown of free fatty acids.
- Complex carbohydrates—fruits, vegetables, whole grains—are good because they convert to blood sugar more slowly and provide energy longer. They are also excellent sources of vitamins, minerals, and fiber.
- Refined carbohydrates—sugar, candy, soft drinks, white bread, etc.—are bad because they have zero nutrition value, they pack a load of calories, and their quick energy jags are rapidly depleted.

Fats

- All fats have nine calories per gram.
- Fats are the main form in which energy is stored.
- They are the energy source for prolonged low-to-moderate intensity exercise.
- Fats are the "shock absorber" for vital organs.
- Fats provide insulation from hot and cold.
- Fats are a source of vitamins A, D, E, and K.
- Simple fats are triglycerides (95 percent of body fat), which are divided into saturated fatty acids, and unsaturated fatty acids. Saturated fatty acids, which are solid at room temperature, are obtained from animal fats and include meat, egg yolks, dairy products, and shellfish. Unsaturated fats, which are liquid at room temperature, come from plant sources, including corn oil, safflower oil, olive oil, and peanut oil.
- Compound fats are simple fats combined with other chemicals. The main ones are phospholipids, which are a key component of cell membranes, and lipoproteins, which transport fat in the blood. These are the HDLs (high density lipoproteins), LDLs (low density lipoproteins), and VLDL (very low density proteins). HDLs, the good cholesterol lipoproteins, ferry cholesterol away from the walls of the arteries to the liver, where it is broken down into bile and excreted through the intestines. The LDLs, the bad lipoproteins, carry the fat throughout the body, where they can form deposits in the arteries, causing atherosclerosis (narrowing of the arteries).
- Derived fats combine simple and compound fats. The best known of the derived fats is cholesterol. Despite its bad reputation, cholesterol is essential for many body functions, including the synthesis of vitamin D, and male and female sex steroid hormones. The liver produces about 500 to 2,000 milligrams of cholesterol a day regardless of how much you take in from food. The main dietary sources of cholesterol are egg yolks, organ meats, shellfish, and dairy products (not the low or no fat kind).

Proteins

- Proteins have four calories per gram.
- *Primus inter pares*: The first among equals, proteins are the foundation of all life.

- Proteins make up one-half of dry body weight, including muscles, skin, bone, hair, teeth, eyes, nails, and scar tissue.
- Hormones and enzymes, which orchestrate all the activities of the body, are proteins.
- Proteins help maintain water and acid base balance, confer resistance to disease, carry oxygen in the blood, maintain growth, and repair of cells and tissue.
- Proteins provide 10 to 15 percent of total energy expenditure during exercise of long duration.
- Protein stores are drawn from muscle and liver for energy when there is not enough carbohydrate or protein available from the daily diet; when people diet without exercising or maintaining adequate physical activity; during extended periods of bed rest or immobilization after injury. This is known as negative nitrogen balance.
- Amino acids are the building blocks of protein. There are two kinds of amino acids, essential and nonessential. Essential amino acid cannot be synthesized by the body and must be supplied from the outside. Nonessential amino acids can be made from nonessential amino acids. Complete proteins are composed of essential amino acids. The absence of even one essential amino acid in the right amount will stop protein synthesis.
- Protein comes from animal and plant sources. Animal proteins—meat, milk, milk products, fish, poultry, and eggs—contain all the essential acids in the correct proportions. Plant proteins—grains and beans—usually lack one or more essential amino acid, but can be combined to form balanced proteins.
- Excess protein is converted to fat and stored in fat cells.

THE FOUR OTHER VITAL COMPONENTS

Foods and liquids supply four other essentials required to operate and maintain a healthy body—vitamins, minerals, fiber, and water.

Vitamins
- Vitamins are organic substances in foods that are essential in small amounts for body processes.
- Vitamins facilitate energy release from fat, carbohydrate, and protein.

- Vitamins are vital for formation of red blood cells, connective tissue, proteins, and DNA.
- Deficiencies cause malfunction of body processes.
- Fat-soluble vitamins are stored in body fat and are not required on a daily basis. There are four—A, D, E, and K. Deficiencies are rare, while excesses of A and D can be toxic.
- Water-soluble vitamins are not stored in fat and deficiencies can occur within two to four weeks. Excesses are eliminated in the urine, making toxicity rare. There are nine—C and B-complex, which includes thiamine (B–1), riboflavin (B–2), niacin, pyridoxine (B–6), cyanocobalamin (B–12), folic acid, pantothenic acid, and biotin.
- Certain vitamins, such as C and E, and beta-carotene, which is metabolized as A, are potent antioxidants, protecting the cells against free-radicals.

Minerals

- Minerals are inorganic elements essential to life.
- Minerals are involved in cellular and energy metabolism.
- Minerals act as enzymes or co-enzymes to regulate chemical reactions.
- Minerals are important for muscle contraction.
- Minerals maintain water balance, acid-base balance, and body fluids.
- Minerals are involved in nerve transmission.
- Minerals are involved in formation of teeth, bone, hemoglobin, protein synthesis, and development of hormones.
- Major minerals have a daily requirement of more than 100 milligrams per day. There are six—calcium, phosphorous, magnesium, potassium, sodium, and chloride.
- Trace minerals are needed in quantities less than 100 milligrams a day and are present in small quantities in the body. There are fourteen—chromium, cobalt, copper, fluorine, iodine, iron, manganese, molybdenum, nickel, selenium, silicon, tin, vanadium, zinc.

While theoretically it is possible to obtain all the vitamins and minerals from a well-balanced diet, as a practical matter, unless you eat your fruits and vegetables straight from your garden, a great deal of the nutrition is lost in the packing, storage, shipping, and cooking of foods. As a result, we advise taking a high-potency daily vitamin and mineral supplement, especially if you are on a weight-reduction program.

Fiber

- Fiber forms the structural wall of plants.
- Fiber is technically not a nutrient since it is not digested.
- Fiber is found in fresh fruits and vegetables, whole grains, nuts, seeds, and legumes.
- Fiber bulks the stools and speeds the removal of waste products through the intestinal tract and out of the body.
- Fiber soaks up fat, fills the stomach, cuts the appetite.
- Fiber may reduce risk of colon cancer and other gut cancers.
- Water-soluble fiber comes mostly from whole wheat products and increases the bulk in the digestive tract for faster elimination.
- Water-insoluble fiber, which includes beans and oat bran, has a high binding activity and is believed to lower serum cholesterol.
- The National Cancer Institute recommends 25 to 40 grams of fiber per day as opposed to the typical American intake of 18 to 20 grams per day. You need both water-soluble and water-insoluble fiber.
- Overuse of bran fiber may result in leaching of calcium, magnesium, and zinc.

Water

- Water is the most essential nutrient for life.
- We can survive weeks without food but only for a few days without water.
- Two-thirds of the body and 85 percent of the brain is water.
- Water is vital to almost every biological process, including digestion, absorption, circulation, and excretion.
- Water is the main constituent of blood and lymph.
- Water regulates body temperature.
- Water lubricates joints and organs.
- Water moisturizes skin.
- Water maintains strong muscle.

- Dehydration, especially in heat and during exercise, causes loss of electrolytes, such as potassium and sodium, and can be life-threatening in advanced stages.
- We need a little over two quarts a day of water, or eight to ten eight-ounce glasses a day. Replenishing water is especially important after exercise.
- Drinking a glass of water before meals cuts the appetite.
- Drink steam-distilled water, for its benefits of absorbing and eliminating toxins from the body.

THE GROWTH HORMONE ENHANCEMENT DIET

Using the nutritional information given above, you can put together your own diet that incorporates all seven elements of the Growth Hormone Enhancement Program. These are

1. Take GH-releasers (see Chapter 16).
2. Bulk up on protein.
3. Keep fats low.
4. Eat carbohydrates that are low on the glycemic index.
5. Keep meals at least four hours apart.
6. Fast one day every two weeks (optional).
7. Do GH-releasing exercises (see Chapter 18).

The important thing in following these guidelines is not how much you eat but how much you eat of each food constituent. Consider the following two tables:

Percent Daily Caloric Intake in Average American Diet

Carbohydrates	40 percent to 50 percent
Fats	40 percent to 45 percent
Protein	20 percent to 25 percent

(The average American diet is even more unbalanced than it appears in the table. Instead of getting 95 percent or more of our carbohydrates from complex carbs, we get fully half in the form of refined sugar and other simple carbohydrates.)

Now here is how most nutritional experts believe we should eat.

Percent Daily Caloric Intake in a Well-Balanced Diet

Carbohydrates	50 percent to 60 percent or more
Fats	20 percent to 30 percent
Protein	15 percent to 20 percent

But with our Growth Hormone Enhancement Program, we are changing the rules. As we said, we want to bring the insulin and growth hormone into balance. This requires yet a different set of percentages.

Percent Daily Caloric Intake in the GH Enhancement Program

Carbohydrates	50 to 65 percent
Fats	10 to 30 percent
Protein	25 to 30 percent

Now, let's look at each of the seven elements in the program.

1. GH-RELEASERS

The key to losing fat and gaining muscle is taking growth hormone releasers. By stimulating higher levels of endogenous growth hormone and IGF–1, you will prime your body to do naturally what it did when you were younger—burn the fat for energy and use the amino acids to build muscle.

Most people should take growth hormone releasers only before going to sleep. This will do two things: give you the best sleep you've had in years and cause the biggest spurts of GH to occur in the first two hours of sleep when they do naturally. If you are exercising heavily, especially if you are doing heavy weight lifting, taking GH releasers an hour (for powders) to an hour and a half (for pills and tablets) before exercising will enhance your strength and performance levels.

2. HIGH PROTEIN

If you've been following the diet controversies that have been swirling around in the last decade, you're probably aware that protein is in the thick of it. Most of us grew up at a time when protein was the hero, lauded for its contribution to strong muscles, bone, teeth, and hair. Then it got slammed for the bad company it kept, especially saturated fats like red meat. More recently, some nutritionists have pointed out that high protein puts a burden on the digestive system and kidneys and have recommended even lower levels of 12 to 15 percent. Now we're telling you to *raise* protein! What's going on?

Okay, here's the skinny on protein. If you're going to sit around and sedentate, stick to the protein levels given above of 15 to 20 percent or even lower. But if you take GH-releasers and exercise vigorously, your body's requirements are going to change.

Brawn Food

Muscles don't appear out of thin air. They are made from the amino acid that the body gets from breaking down protein. If not enough protein is supplied from the outside in the form of food or nutritional supplements, the body draws on its own tissue, in effect eating its own muscle. This is muscle-wasting. When this occurs the body is in a state of negative nitrogen balance. That means it does not have enough nitrogen, which comes from the amino acids, for tissue-building. Some athletes use anabolic steroids to keep themselves in positive nitrogen balance, which means that they are retaining more nitrogen, and thus more protein building blocks for muscle growth. This is the anabolic state versus the catabolic state, in which more tissue is being broken down than built up.

But anabolic steroids are a dangerous and unnatural way to bulk up muscles. The right way to do this is by increasing the percentage of protein you eat in your diet and replacing your growth hormone levels with GH-releasers.

Effect of GH on Protein

Remember what growth hormone does to lean body mass—it builds it up by increasing protein synthesis. And it puts this protein into the lean body mass, including muscles. Ordinarily your body likes to converts some of the calories in protein to sugar and then changes that sugar into fat, which it stores for future use. But growth hormone performs this alchemy trick where it uses the protein for muscle-building and mobilizes the stored fat so that it can be used as fuel. This means that you have to feed the protein-making machinery with more protein, so that you don't get into negative nitrogen balance, where you are digesting your own muscles to get the needed protein.

How Much Protein Do You Need?

We can get some idea of the protein needed for a GH-enhancement diet from studies of elite athletes. According to a 1984 article by Peter Lemon and co-workers in *Sports Medicine*, "Athletes should consume 1.8 to 2.0 grams of protein per kilogram of body weight per day [or .8 to .9 grams per pound]. This is approximately twice the recommended requirement for sedentary individuals."

Other researchers suggest even higher percentages. A study of elite weight lifters in Rumania found that when they increased their dietary program to 1 to 2 grams per pound of body weight (225 percent to 438 percent of the U.S. Recommended Dietary Allowance), they increased their muscle mass about 5 percent and their strength performance by about 5 percent. Russian researchers who experimented with high-protein diets and vigorous weight lifting found that during periods of intensive training, the athletes required about 2.2 to 2.6 grams per kilogram of body weight, or about 1 to 1.2 grams per pound. Some athletes even went into negative nitrogen balance when they went below 2 grams per kilogram during intensive training.

We recommend that you get anywhere from .5 to 2 or even more grams per kilogram of body weight. If you are a diabetic, have kidney problems, are sedentary, or are over age sixty-five, it is best to stay on the low side. However, many of you will want to increase your exercise levels, including resistance weight training. Depending on your level, you can go between 1 and 2 grams, or even higher. It is important that you drink between 10 and 12 glasses of water to help the kidneys flush out toxins.

The indicators for increased protein are increases in muscle mass, strength, and body weight. But to do this scientifically, you should have tests of body percentage fat done by your doctor, such as ultrasound, infrared, or electrical impedance. It is also vital to maintain proper nutrition using the information given above.

Where Should the Protein Come From?

The best sources of dietary protein are lean chicken, fish, occasional lean meat, soy products, egg whites, and whey protein. Some bodybuilders swear that they get bigger muscles when they eat chicken and turkey instead of red meat. And who are we to argue with bodybuilders? But red meat also contains carnitine, a metabolic enhancer (see Chapter 18).

Then there is the wonderful, versatile soy protein. Tofu, or "to-faux" as we like to call it, can be made to resemble almost everything under the sun from chicken salad to hamburgers to ice cream. Orthodox Jews who always secretly craved cheeseburgers can now indulge in real hamburgers made with "to-faux" cheese. Soy protein contains a powerful antioxidant, genistein, and a protease inhibitor that provide a double whammy against cancer. It also boosts HDL levels, lowers triglycerides, and maintains blood glucose at healthy levels. The best sources of soy protein are soy flour, soy milk, tofu, soy nuts, miso, and tempeh. *Not* soy sauce or soybean oil, which contain very little of the protein.

Soy protein has yet another remarkable effect that will light up the life of anyone trying to lose weight. It naturally boosts the levels of thyroid hormone. Thyroid hormone is a major culprit in the devastating diet phenomenon known as the rebound effect. No doubt, you've experienced this one for yourself. You try your darnedest to lose weight, struggling to get off every pound, only to have them bounce back with a vengeance the moment you stop dieting. That's because your thyroid hormone level actually decreases during a diet, causing your body's metabolic rate to slow down. It's part of the starvation reflex we talked about above. Soy protein helps counteract this effect by keeping levels of thyroid hormone high while you're dieting and afterward. A new highly concentrated extract of soy protein is available through the Life Extension Foundation in either tablets or powder; it has a pleasant taste resembling peanut butter and can be spread on bread. (See Appendix for suppliers.)

Egg whites are pure protein. You can make egg whites from hard-boiled eggs palatable by chopping them up with low-fat mayonnaise and a little salt. Whey protein is another favorite of bodybuilders. Whey is the runny stuff that's left when you make cheese.

The easiest way to get your protein, and the one we recommend if you are going in for the whole program, is a high-quality protein shake. Here is one recipe:

PROTEIN EGG SHAKE
2 to 6 boiled eggs with the yolks removed
1 to 2 bananas
Half cup strawberries
1 teaspoon vanilla
Mix in blender at high speed until well blended.

You can vary your shakes by adding skim milk or soy protein or even vegetables like broccoli. You can also buy commercially made protein shakes at the health food store and add water, milk, juice, or fruit, and blend it, according to the instructions. If you are using protein shakes to meet your protein requirements, take one or two protein shakes a day, the first between breakfast and lunch, and the second between lunch and dinner or immediately after exercising.

3. LOW FAT

Fat is the enemy of growth hormone. It blocks both the production and the release of GH. "If you're eating 40% to 50% every day and you're taking growth hormone, it's like putting on the heat and the air conditioner at the same time," says Dr. Terry. We give you a range of up to 30 percent because many people find it difficult to go below this. But for best results, aim for the low end, with most of this coming from unsaturated fats or monounsaturates like olive oil. While growth hormone raises HDL levels and lowers LDL levels, there's no sense in working against this action by eating saturated fats that increase cholesterol.

4. EAT LOW-GLYCEMIC CARBOHYDRATES

The glycemic index is a measure of how fast a carbohydrate enters the bloodstream and raises your blood sugar level. The index forms a basic part of the GH diet because when blood sugar is raised, it triggers the release of insulin. The higher the food is on the glycemic index, the more insulin is secreted. And high insulin blocks the fat-melting action of growth hormone just as growth hormone interferes with the fat-storage action of insulin.

Many of the carbohydrates you would expect to be low on the glycemic index are, such as most vegetables. But then there are surprising exceptions like carrots, corn, and beets, which are all very high. Most fruits, even very sweet ones like berries, are low, because their fiber content slows the entrance into the bloodstream and because fructose, the sugar in fruits, has a slower entry rate than glucose.

This brings us to the most surprising and least intuitive carbohydrates that are high on the index. Glucose is the sugar found in grains and starches like bread, cereal, potatoes, rice, and pasta. These are the very foods that have been touted as the basis of the new healthier, low-fat way to eat. But these are also the foods that raise the level of insulin, which encourages the storage of fat. In fact, some nutritional experts believe that the glycemic index helps explain why Americans are getting fatter while they claim to eat less fat than ever before. As a nation, we *are* eating less fat, especially saturated fat, which helps account for the significant drop in the rate of heart disease. But the pasta and grains we conscientiously substitute for meat have raised insulin levels, generated weight gain, and caused the rate of insulin-resistant, type 2 diabetes to rise dramatically!

This does not mean that you should cut out all high-glycemic foods. Almost everything, including a dish of Häagen-Dazs, is all right if done in moderation. Incidentally, ice cream is low on the glycemic index and low-fat ice cream or frozen yogurt is even better. But it does mean that you should familiarize yourself with the glycemic index (see table below) and try to eat foods that are on the low end rather than on the high end. This means, for instance, eating sweet potatoes rather than regular potatoes and whole wheat spaghetti rather than white rice. Adding fiber is a good way to retard the rate of insulin's entry into the bloodstream. High-fiber oatmeal, for instance, has a glycemic index of 49, while cornflakes are 80.

TABLE OF GLYCEMIC INDEX, PROTEIN, FAT, AND CARBOHYDRATES

Food	Glycemic index	Percent Protein	Percent Fat	Percent Carbohydrates
Bakery goods				
Pastry	59			
Sponge cake	46			
White bread	69	9	3	51
Whole wheat bread	72	9	3	49
Whole-grain rye bread	42			
Dairy Products				
Ice cream	36	5	11	21
Milk, skim	32	4	below 1	5
Milk, whole	34	4	4	5
Yogurt	36	3	2	5

TABLE OF GLYCEMIC INDEX, PROTEIN, FAT, AND CARBOHYDRATES

Food	Glycemic index	Percent Protein	Percent Fat	Percent Carbohydrates
Fish				
Fish sticks	38	17	9	6
Fruit				
Apples, Golden Delicious	39			
Bananas	62	1	below 1	22
Cherries	23			
Grapefruit	26	below 1	below 1	4
Grapes	45	1	below 1	17
Orange juice	46	1	below 1	10
Oranges	40	1	below 1	12
Peaches	29	1	below 1	8
Pears	34	1	below 1	14
Plums	25	1	below 1	19
Raisins	64	2	below 1	77
Grains				
All-Bran	51			
Brown rice	66	1	1	25
Buckwheat	54			
Cornflakes	80	8	below 1	85
Oatmeal	49	2	1	10
Shredded wheat	67			
Swiss muesli	66			
White rice	72	2	1	24
Spaghetti	50			
Whole wheat spaghetti	42			
Sweet corn	59	3	1	19
Meat				
Sausages	28			
Nuts				
Peanuts	13	26	49	21
Sugar				
Fructose	20			100
Glucose	100			100
Honey	87	below 1		82
Maltose	110			100
Sucrose	100			

TABLE OF GLYCEMIC INDEX, PROTEIN, FAT, AND CARBOHYDRATES

Food	Glycemic index	Percent Protein	Percent Fat	Percent Carbohydrates
Vegetables				
Baked beans, canned	40			
Beets	64	2	below 1	10
Black-eyed peas	33			
Carrots	92	1	below 1	10
Chickpeas	36			
Kidney beans	29	8	1	21
Lentils	29	8	trace	19
Lima beans	36	8	1	20
Parsnips	97	1	1	15
Peas, frozen	51	5	below 1	12
Potato chips	51	6	40	50
Potato, instant mashed	80	2	3	14
Potato, russet, baked	98	2	below 1	16
Potato, sweet	48	1	below 1	19
Potato, white	70	2	below 1	13
Soybeans	15	11	6	11
Tomato soup	38	2	2	13
Yams	51	2	below 1	20

5. SPACE YOUR MEALS

Take advantage of the feast-fast cycle (see above) by spacing your meals and snacks so that they are between four hours and five hours apart. This time interval will allow the insulin to disappear from your bloodstream so that the growth hormone can work unimpeded to build muscle and lean body tissue. But don't go beyond five hours or you will suffer from the effects of hypoglycemia. The slower-absorbed proteins in your diet should keep you from becoming hungry. Have your last meal at least four hours before you plan to go to bed. When you go to bed at night, you'll be taking your growth hormone releasers on an empty stomach so that they can be fully absorbed.

6. TWENTY-FOUR-HOUR FAST

As we mentioned, the greatest spikes of growth hormone occur during fasting. In fact, the anti-aging effects of growth hormone may help explain why caloric restriction, or fasting every other day, has doubled the life span of experimental animals. It is probably one of the body's survival mechanisms in times of food shortage, maintaining stamina and strength. People who are in good health can try fasting for one day every two weeks to really get their growth hormone levels surging. Roy Walford, M.D., one of the nation's pre-eminent gerontologists and author of the *120-Year Diet*, has used fasting for two days of every week for years as an expedient way to cut down calories for life-extension purposes.

Since you will not be taking in energy in the form of food, plan to fast when the physical and mental demands on your time are at their lowest. For most people, not eating for twenty-four hours makes it difficult to perform vigorous exercise or other taxing activities. To prepare your body for a fast, a day or two before, switch to a simple diet of cooked vegetables, salads, and juices, while avoiding grains, breads, dairy products, meat, and fish. Another pre-fast approach is to skip a meal or two for a few days before beginning the fast. Fasting for a single day is not very difficult to do, and many people report that they experience a euphoric feeling similar to a runner's high.

In addition to hiking GH levels, fasting allows the digestive enzymes to move through the empty stomach directly into the bloodstream and intestines, cleanses the bloodstream of metabolic wastes, purifies the colon, and relieves allergies, headaches, inflammations, blood pressure problems, and various skin diseases. Important: Do not attempt to fast without consulting your physician, especially if you suffer from diabetes, heart disease, or other medical conditions.

When fasting, here are a few simple rules to follow:

1. Drink two to three quarts of liquid, or eight to twelve glasses. This can be in the form of fresh vegetable or fruit juice, divided up during the day, with one to two glasses in the morning as breakfast; one glass at midday; one to two glasses for lunch; and one to two glasses for dinner. The rest of the time, drink water.
2. Do not maintain a busy schedule this day. Relax, do light exercise, and nap when necessary. You shouldn't really need as much sleep as usual, since your body doesn't have to work to digest food.

7. GH-RELEASING EXERCISES

Both aerobic exercise at very high levels and weight training will stimulate growth hormone. In the next chapter, we'll give you the workout plan that together with growth hormone releasers and the GH diet will take decades off your body.

Maximizing the Effect: Exercise and Strength

You don't have to exercise to get the benefits of growth hormone. But raising growth hormone levels without exercising is like driving a Mercedes on regular gas instead of high octane fuel—it works but you won't get great performance. Growth hormone releasers will still melt away fat while building muscle. Growth hormone and exercise are like caviar and champagne; they bring out the best in each other. Stimulation of growth hormone enhances your capacity for intensive exercise, while strenuous workouts increase growth hormone levels. Together they can reverse the downward spiral that starts after age thirty-five, when, like Alice in Wonderland, you have to run just to stay in place. In this chapter, we'll give you the exercises that will make you look, feel, and perform like a champion.

IS GH THE SECRET OF OLD BODYBUILDERS?

Meet Bob Delmonteque, as fine a specimen of manhood as you'll ever see— bulging biceps, massive pecs, satiny skin, flat abs, slim waist. Sneak up on Bob from the back and except for his thick gray mane, you'd think he was a twenty-nine-year-old. Actually, he's probably the only seventy-six-year-old who looks better now than he did at nineteen. He's one of the nation's foremost fitness experts, a top model for Weider Publications, and a trainer who's built the bodies of the original seven astronauts and such Hollywood legends as Clark Gable, Errol Flynn, John Wayne, and, coming into the 1990s, Matt Dillon. Bob's not on growth hormone. He keeps his HGH levels in the optimal range by vigorous exercise, proper diet, rest and recuperation, positive thinking, and a daily helping of natural HGH-releasers. "You really can activate the pituitary yourself," he says.

His workout would daunt a much younger man. On Monday, for instance, he does his chest, a series of twelve exercises, three sets, eight to ten reps. And he exercises to exhaustion for an hour and fifteen minutes with a pulse rate that can go to 170. "I don't even get a training effect," he says. He still bench presses 250 pounds, runs marathons in five hours and twenty-three minutes, goes on 100-mile bike rides, swims out in the ocean for two miles and does one hundred sit-ups before going to bed every night. "I have grandkids and they can't keep with me. I'm in superb shape."

Delmonteque believes the mental and spiritual aspects are just as important as the nutritional aspects. "The brain is an apothecary," he says. "And it is the strongest drug you can purchase. I have faith, belief, enthusiasm, hope, a blazing determination, a will to do it, and I persist until I succeed."

Sexually, he's in even better shape (if that's possible). He tells the story of being approached several minutes earlier by a young woman of twenty-three, who said, "'Man, you look so good. I could take you to my room, right now.' Being a married man, I declined, but the reality is if you don't use it, you are going to lose it. That's my theory—body, mind, spirit, and sex."

Delmonteque's workouts are a primary cause of his elevated IGF–1 levels. The endorphins he's creating with his intensive exercise and mental outlook probably help too. But the anabolic effect of pumping iron, which causes high peaks of GH, may well be the secret to such ageless wonders as fitness magazine magnate Joe Weider (age seventy-two); health club and exercise guru Jack La Lanne (age eighty-two); winner of the Master's Mr. Olympia, Ed Carvey (age sixty-two); and Helen Zechmeister (age eighty-one), who holds the unofficial power-lifting records for her age group (there are no classifications for senior citizens), including deadlifting 245 pounds!

NO PAIN, NO GAIN IN GH LEVELS

You have to go for the burn to get those all-important growth hormone spikes. (It is the bursts of GH that get the hormonal messages to all the cells of the body.) The study that showed this involved twenty-one normal healthy women between the ages of eighteen and forty, who were not in an exercise program. The women were divided into two training groups who ran at moderate and high levels of intensity, and a group of nonrunning controls. Although both groups covered the same mileage per week, the high-powered group trained six times a week as against three times a week for the other group. And they also differed in how hard they ran as measured by their lactate threshold (LT).

The lactate threshold is the point at which the buildup of lactic acid in the muscles causes fatigue and strain. It is different for each individual as measured by blood levels of lactate. The high intensity group ran at their LT, that is, pushing themselves to the limit, while the moderate intensity group ran below their individual LT.

At the end of the year, when both training groups were running thirty-five to forty miles a week, according to Arthur Weltman and his associates at the University of Virginia in Charlottesville, the high-intensity runners had much larger growth hormone peaks compared with their measurements before they started running, but there was no change in GH release in either the moderate-intensity group or the controls. Running at top intensity, note the researchers, also pushes up the level of beta endorphin in the blood, the "runner's high," which has been shown to stimulate GH secretion. This kind of endurance training, modified to suit the individual, may also raise growth hormone levels in both obese people and the elderly, they say, with all the associated benefits, such as improved body composition and well-being. According to the researchers, both groups of intense exercisers significantly increased their lean body mass as result of their year of training.

GAIN WITHOUT PAIN

While pushing the exercise envelope may jack up growth hormone levels, replacing growth hormone in people who are deficient increases their ability to do exercise without increasing their effort!

In a double-blind study, Peter Sönksen and his associates at St. Thomas' Hospital in London treated GH-deficient adults for six months with growth hormone injections, while a control group received placebo shots. The GH-treated group had all the favorable changes in body composition, such as loss of fat and increased muscle mass and strength (see Chapter 9), but they also were able to increase their exercise performance. Their VO2 max, or maximum oxygen uptake, a measurement of lung capacity and endurance of the cardiovascular system, increased 17 percent, compared with that of the control group. This rise was "paralleled by an increase in maximal power output," note the researchers, and in anaerobic threshold (the point at which you're gasping for air because the cells of your body are using more oxygen than your respiration can provide).

Amazingly, during the period of maximum exercise, there was no difference between the GH-treated group and the controls in measurements of effort, such as maximum or resting heart rate, or in the subjective experience of exertion. Apparently, the increase in growth hormone levels, along with the changes in body composition, lifted them onto a higher level of strength and performance. Because the rise in exercise performance was even more dramatic than the modest increase in strength they were able to measure, the researchers believe that the growth hormone treatment enhanced the type 1 muscle fibers over the type 2 muscle fibers. (Type 1 are the white muscle, or fast twitch, fibers, involved in powerlifting, while type 2 are red muscle, or slow twitch, fibers increased by running.)

The ability to do more work without more effort has great significance in daily life. The researchers note that "if a given task can be done with less physiological stress, the patients may be inclined to increase their work productivity. Indeed, many of the patients spontaneously described an increase in the ease of completing daily tasks."

GOING FOR THE GOLD

One double-blind study of elderly men on growth hormone replacement, conducted by Maxine Papadakis, M.D., and her associates at the University of California in San Francisco, did not find increased muscle strength. But L. Cass Terry, M.D., and Edmund Chein, M.D., in their analysis of 202 patients who have been on growth hormone for more than six months, found "definite increases in strength, muscle mass, and decreased fat" (see Chapter 3 for details of study). In addition to the self-assessment used in their study, Terry has personally interviewed a large number of the patients. "Every one I've talked to, some of whom are physicians, have said that they had marked increases in energy, muscle strength, stamina, and endurance," he says.

The difference in results between Chein's patients and those studied by Dr. Papadakis, Terry believes, is a matter of dosing. While the San Francisco team used doses that were far higher and gave them three times a week, Chein and Terry use low doses twice a day that are individualized for each patient on a six-days-a-week basis (see Chapter 19 for optimum dosing). "She gave large doses that were not physiologic, so there were side effects. It didn't even come close to mimicking the natural pituitary secretion of growth hormone. With our study, there has been almost zero side effects." Interestingly, the lower doses Chein and Terry used did not translate into lower IGF–1 levels, but rather higher levels that could account for the increased benefits. While the San Francisco study maintained IGF–1 levels between 150 and 350 nanograms per milliliter, Chein tries to maintain his patients between 300 and 350, which is the level of people between twenty and thirty years old. "If you're running about 190 which is where many of the patients in their study were at, you're in the 50 to 60 year-old age group," says Terry. "That's too low. Either their livers weren't responding to the growth hormone and making IGF–1 or they weren't dosed properly."

Another possibility is that the higher GH doses could have turned down the levels of IGF–1 in the San Francisco study. This is a major reason only replacement doses of growth hormone and other hormones should be used, so that you don't shut down the secretory capacity of the gland. For instance, this has been documented with corticosteroids, where overuse has caused the adrenal glands to stop producing.

Terry also points out that the patients in the Papadakis study did not combine growth hormone with exercise. "I believe that you can't take hormones and sit on your duff," he says. "Anyone who has 13% increase in muscle mass has to have more strength if they exercise. Both aerobic and resistance training are part of the age-deterrent system."

According to Chein, there is no doubt that his patients have greater strength and endurance. "I have 50 and 60-year-old patients going into the gym and the 20-year-olds can't beat them in weights or in the speed in which they gain muscle mass," says Chein. "Some of them are bench-pressing 250 pounds one week, 260 the next week, 270 the following week, going up as much as ten pounds a week."

For those who routinely measure their performance, there is a noticeable difference. R.T., the forty-eight-year-old lawyer, cut two minutes off his three-mile-a-day walking program and Paul Bernstein, Mr. Fitness, upped the output of his nightly workouts by 20 percent with growth hormone. A patient of Dr. Baxas, who began a workout program, advanced three times faster in a six-month period than his friend of the same age who was not using GH. And like the GH-deficient patients in the St. Thomas' Hospital study, the people on growth hormone replacement are getting the gain without an increase in pain. "Growth hormone produces increases in strength, tendons, and joints," says Ahlschier, who has been on it for more than two years. "There is a tremendous increase in muscle tone, absolutely phenomenal without doing any exercise. You can flex your muscles and you won't believe how hard they are."

But many physician-users of growth hormone and GH stimulators find that the combination of exercise and hormone replacement is pushing their middle-aged bodies to Herculean heights. Ahlschier, who has "religiously and fanatically" worked out for the past two years, says "I don't know many 55-year-olds who look like I do." At 267 pounds, he is all muscle and can do leg presses of over 1,000 pounds. Another advantage of growth hormone, he points out, is that it protects the body against cortisone, the stress hormone, that is produced during exercise. "Cortisone is the enemy we fight everyday with its catabolic effect tearing down our bodies. If you work out heavily and take growth hormone, you have less catabolic effect from cortisone. You grow faster, do less damage to your tissue, and repair quicker."

Terry, a self-described "health freak and fanatical weight lifter," who also does a lot of aerobic exercise, found that growth hormone "made a tremendous difference in my fat composition and body configuration in less than two months. It increased my stamina and my strength in a more rapid rate, I think, than would have occurred if I had not been taking it." After five months on GH therapy, he says, "I don't get nearly as fatigued as I did. The dumbbell bench press on which I did 30 pounds, I am now up to 50 pounds and I can kill it."

BENEFITS OF EXERCISE

Exercise is one of the greatest anti-aging bullets that is available to everyone. A list of the benefits of exercise resembles that obtained with growth hormone: muscle growth and strength, loss of fat, increased energy, greater well-being, and decreased anxiety and depression. It raises the levels of HDL cholesterol, lowers blood pressure, improves the immune system, and helps protect the body against a host of diseases, including cardiovascular disease, stroke, hypertension, diabetes, and osteoporosis.

EXERCISE INCREASES LONGEVITY

Moderate exercise extends your life. A famous study looked at almost 17,000 male alumni of Harvard University who were between the ages of thirty-five and seventy-four. As the physical activity of the men increased, their death rates decreased, according to Dr. Ralph Paffenbarger and his colleagues at Harvard and Stanford universities. Men who expended at least 2,000 kilocalories doing such moderately vigorous exercise as playing tennis, swimming, jogging, or brisk walking lowered their overall death rates by 25 to 33 percent and decreased their risk of coronary artery disease by an astounding 41 percent when compared with their more inactive fellow alumni. Going over 3,500 kilocalories actually made things worse, with a slight increase in the death rate. The researchers calculated that by the age of eighty, the exercisers had added one to two years to their life span compared to the nonexercisers.

EXERCISE STRENGTHENS THE OLDEST OLD

A 1994 study led by Maria Fiatarone, M.D., of the Human Nutrition Research Center on Aging of Tufts University in Boston gave a multinutrient supplement and exercise training to a group of men and women in a nursing home between the ages of sixty-three and ninety-eight. Even though 83 percent of them required a cane, walker, or wheelchair and 66 percent had fallen during the past year, they all went through high-intensity, progressive weight-training of the hip and knees for forty-five minutes, including leg presses, three times a week.

In only ten weeks, the results were remarkable. The muscles of these elderly people grew in size and their strength increased, with the greatest benefit seen in those who were weakest to begin with but who did not have severe muscle atrophy. They were also significantly more mobile than before the study, walking faster, climbing stairs more easily, and doing more physical activity. Four of the exercisers graduated from a walker to a cane, compared with one person in the group of nonexercising controls, who went from a cane to a walker during that time.

EXERCISE AND GH

Exercise sends a wake-up call to your pituitary. Just getting started in a running program or weight lifting will stimulate your pituitary gland to secrete higher growth hormone levels. "Exercise and decreased food intake is the perfect prescription for the U.S. population to increase GH," says Bengtsson. "We have totally stopped moving. The public enemy number one for health is television. We can without any doubts, without any expenses, increase our growth hormone secretion just by moving. Why does your bone mass increase with exercise? Possibly because your growth hormone increases."

The way exercise works to stimulate growth hormone is not understood, according to Stephen Borst and his associates at the University of Florida. They suggest such possible mediators in the body as low blood sugar, lactate accumulation, and release of beta endorphins.

EXERCISES THAT RAISE GROWTH HORMONE RELEASE

There are two forms of exercise that stimulate growth hormone and they do it in two different ways. *Aerobic exercise* results in persistent long-term release of growth hormone in the plasma for two hours or even longer after you stop exercising. *Resistance exercise* like weight-training causes the spurts of growth hormone that stimulate IGF–1.

Moderate intensity rather than high intensity may be enough to stimulate growth hormone release. "Aerobic exercise of moderate intensity may be sufficient to cause maximal stimulation of GH release," according to Stephen Borst, William Millard, and David Lowenthal at the University of Florida in Gainesville. Moderate intensity of aerobic exercises is about 40–50 percent of maximum oxygen uptake while high intensity is 70 percent.

Type of Exercise	Intensity	GH Effect
Stationary bike	Moderate	145 percent increase
	High	166 percent increase
Treadmill	High	Enhanced pulsatile secretion
Running (men)	Moderate	None to moderate
Running (women)	High	Elevated IGF–1
		266 percent increase in lowest GH level
		75 percent increase in twenty-four-hour serum GH
Weight training	High	Rapid and sustained increase in both men and women
	85 percent of maximal lift capacity	Four-fold increase
	70 percent of max	Three-fold increase

CHOOSING A GH-RELEASING EXERCISE

Using the table above as a guide, you can choose an aerobic exercise to enhance GH plasma levels. This includes running, race-walking, treadmill running, and stationary bicycle. Follow these rules when exercising:

1. Start at low intensity and build up slowly.
2. Exercise for thirty minutes.
3. Exercise three to six times a week.
4. If you are over forty and have been previously sedentary, aim for moderate intensity in the beginning. Once you are conditioned,

you can start trying for higher intensity levels of 60, 70, or even 80 percent of maximum oxygen uptake. (See formula below for how to figure your maximum uptake using your pulse.)

Caution

Do not attempt to start any exercise program without first obtaining a comprehensive sports medicine physical by a knowledgeable physician.

WALK-JOG-RUN

Two levels of running promote the release of growth hormone. The first is running as the quickest way to lose fat and change your body composition in a more favorable direction. As you gain lean body mass, your body becomes more metabolically active and releases more growth hormone. The second level is running to the point where the activity itself is growth hormone–releasing. That requires a high level of conditioning. On either level, running or race-walking does great things for the body. It has the highest calorie burn per minute of almost any exercise or sport you can do, including football, cycling, squash, or rowing. It increases aerobic fitness, raises the HDL levels, stimulates the mechanisms that detoxify the body, opens the microcirculation in your lungs, brain, and all the other organs of your body, and it tones the neuroendocrine system.

Running guru Kenneth Cooper, M.D., reports a study by his research group at the Cooper Aerobics Center in Dallas in which three groups of premenopausal women were trained to walk three miles a day, three days a week, for six months at various speeds. A fourth, nonwalking group served as the controls. The first group, which walked a modest twenty minutes per mile, increased their aerobic fitness by 4 percent; the second group, which walked a brisk fifteen-minute mile, raised their aerobic fitness by 9 percent; and the third group, which race-walked twelve-minute miles, jumped an astounding 14 percent in aerobic fitness, more than triple that of the slowest walkers. In fact, their energy expenditure and heart rate were equivalent to those of women who ran nine-minute miles. Both weight loss and aerobic fitness are dose-related; the faster you go, the more fuel you use, the better your heart and your metabolism respond.

If you are overweight and just starting an exercise program, it is essential that you see your doctor and have a stress electrocardiogram if you are a woman over fifty or a man over forty. Anyone over sixty should have cardiovascular screening as well as a stress EKG. In the same way that you have to crawl before you walk, you have to walk before you run. Cooper advises that you start by taking the following steps:

1. Keep track of your distance by measuring the mileage you expect to cover.
2. Warm up with three to five minutes of calisthenics or jogging.
3. Cool down with by walking for five minutes after you've gone the distance.

Start out by walking a half hour a day, three to five times a week. Your goal is get your heart rate into the aerobic fitness range, which is based on a simple formula: 220 minus your age times 70 percent. Thus, if you are fifty years old, your target range would be 220 minus 50 times 70 percent, which equals 119 beats a minute. Immediately on finishing your run and before starting your cool down, check your pulse for fifteen seconds and multiply by four to get your pulse rate per minute.

You can incrementally increase the distance, speed, and duration keeping your heart in the target range. Race-walking is practically as good as running and it has the advantage of causing far less wear and tear on the body.

Another way to get into running that some people find congenial is to combine walking and running from the start. If you like, you can vary the routes to keep up your interest. As the mood hits, break into a run every now and run for a short while. Let yourself go and feel the joy of running. Remember what it was like when you were a child and you ran for the sheer fun of it, because you liked the way it felt when your legs moved, and you knew if you really wanted to, you could fly. When you tire, drop back into a walk. As you walk-run, you'll find that the ratio between the two changes, until you're running more than you walk. You have to run or race-walk long enough and hard enough to get into the fat-breakdown phase of exercise. In the first ten to twenty minutes, you use up your first fuel for energy, glycogen—the carbohydrate reserve stored primarily in the liver and muscle. After that the body has to draw on its fat stores. Once you start burning fat, you turn on the lipase enzymes, which break down the fat cells. This fat metabolism continues even after you stop exercising!

HIGH-INTENSITY RUNNING

Strenuous endurance running may be needed to restore those youthful GH pulses and stimulate the IGF–1, which has most of the important benefits. Vigorous exercise creates lactic acid, which benefits the heart, as well as binds and removes toxic metals that accumulate in our bodies. The lactate threshold is the point at which you feel muscle fatigue and strain.

Running at the lactate threshold puts you into the elite level of runners. In Kenneth Cooper's book *It's Better to Believe*, he lists the goals of elite race-walkers for those under forty-nine as eleven minutes and thirty seconds per mile; for ages fifty to fifty-nine, it is the same; for those sixty to sixty-nine, it is thirteen minutes per mile; and for those over seventy, it is fifteen minutes per mile. These levels are all attained after ten weeks of training by people who are competitive aerobic walkers or race-walkers to begin with. Since lactate tolerance has to be individualized, anyone interested in running or race-walking at this level has to work with a doctor or personal trainer.

TONING THE NEUROENDOCRINE SYSTEM

The neuroendocrine system is the biochemical connection between the brain and the body. Running, race walking, and aerobic exercise actually stimulate the NE system, releasing the brain chemicals that work on the body's cells and tissues. Recent studies show that the release of beta endorphin (the so-called runner's high) and the catecholamines, such as dopamine, norepinephrine, and epinephrine, are related to the intensity of exercise and the lactate levels. All these brain chemicals stimulate growth hormone secretion. But the workout has to be of very high intensity, since exercise at 60 percent of V02 max (maximum oxygen uptake) had no effect on endorphin levels, while upping it to 70 to 80 percent of V02 max markedly raised the circulating levels of endorphin. In other words, intensive exercise is equivalent to taking a potent growth hormone releaser!

MUSCLING YOUR WAY TO GROWTH HORMONE

The best exercise, bar none for sending those GH levels surging is weight-resistance training. In combination with growth hormone replacement or GH-releasers, lifting weights can do things to your body that you never dreamed possible. The exercises below will allow you to get deep cuts—bodybuilder parlance for those stand-out, bulging muscles. Men, here's how to look "ripped to shreds" like one of those bodybuilding gods you've always secretly admired. Women, you can have a well-toned, shapely body with strong pectorals providing a natural uplift to your breasts. And if you want to look like a rippled, sexy Amazon, you can do that too.

LIFTING WEIGHTS AT ANY AGE

How do you know if you're too old to start lifting weights? Try this exercise devised by George Burns, AKA God. One, put your hands up over your head. Two, put your hand in front of your chest and push out as if you were doing a pushup. Are your hands touching anything? No? Then you're not in your coffin and you're still able to work out.

All workouts have three components, intensity, frequency, and duration. Intensity is the level of effort, frequency is how often you work out, and duration is how long a single workout lasts. For example, an elite athlete may lift 200 pounds for three repetitions on an almost daily basis, while a sixty-five-year-old may only be able to handle sixty pounds for three reps every other day. And while the elite athlete may train for two hours, the older person may work out for half an hour. The intensity, frequency, and duration for the two individuals may differ, but the workout is still the same. They are both doing the same lifts, using the same biomechanics, at weights that stress their bodies to the same degree.

STARTING OUT

After you've gotten past the couch potato stage, you are ready to begin weight-resistance training. Start each exercise workout, whether it is aerobic or weight training, with a warm-up. Here is a basic series of six warm-up exercises. The amount of repetitions is given for each exercise for beginners (B) who are out of shape and intermediates (I) who work out on a fairly regular basis. Beginners and intermediates should progress slowly to the next level. Other warm-ups should be added as your training advances.

1:   *Bend and stretch* (lower back and hamstrings)

Stand erect, feet shoulder-width apart, hands on hips. Exhale and bend trunk forward and down and allow arms to hang down in front of you. Flex knees slightly as you reach toward toes. Pause for a count of three and gently try to straighten knees to a locked position. Return to starting position and inhale. Reps: B—five reps, rest twenty seconds, then repeat once. I—fifteen reps, rest twenty seconds, and repeat.

2:   *Knee lift* (legs, knee joints, back, shoulders)

Stand erect, head up, feet shoulder-width apart, hand on hips. Raise knee up toward chest as high as possible and bring arms forward to lock around lower leg below the knee joint. Exhale and pull against arms for a count of three. Return to the starting position and inhale. Repeat exercise with opposite leg. Reps: B—five reps, rest twenty seconds, repeat one more set; I—fifteen reps each side, rest twenty seconds, repeat.

3:   *Half-knee bend* (thighs, knee joint, Achilles tendon)

Stand erect, head up, feet shoulder-width apart, hands on hips. Inhale and bend knees halfway while extending arms forward, palms down. Pause for count of three and return to starting position. Exhale. B—ten reps each side, rest twenty seconds, then repeat one set. I—ten reps, rest twenty seconds, repeat one more set.

4:   *Arm circles* (shoulders, upper back, arms)

Stand erect, head up, feet shoulder-width apart, hands on hips. Extend arms straight out to sides. Rotating arms at shoulders, describe small circles—fifteen backward circles, then fifteen forward circles. B—one set. I—twenty circles forward and back; rest fifteen seconds, repeat.

5:   *Body bender* (rib muscles, lower back, and hip flexors)

Stand erect with hands locked behind neck, feet shoulder-width apart. Inhale and bend trunk sideways to right as far as possible.

Return to starting position and exhale. Repeat movement to the left. One complete right/left movement counts as one repetition. B— bend to each side ten times, rest fifteen seconds, and repeat. I— twenty reps each side, rest fifteen seconds and repeat.

6:   *Knee pushup* (shoulders, arms, back)

Lie facedown on floor with legs together, knees bent with feet raised at right angle to floor. Hands on floor under shoulders. Inhale, then exhale as you push upper body off floor until arms are fully extended and upper body is in straight line from head to knees. Return to starting position and inhale. B—add knee pushup to warm-up after one week of workout program, three to six reps, rest twenty seconds, and repeat. I—ten reps, rest twenty seconds, repeat.

HANDLING THE WEIGHTS

1:   Initially, stick to light weights at a resistance you can handle doing five to eight repetitions.

2:   Use slow controlled movements so that your musculature is controlling the weight as you go through the range of motion. (You may have heard about taking four seconds to lift the weight and four seconds to bring it down, but this rule was invented only because some people moved the weights on the Nautilus Machine so slowly that they actually broke the cables.)

3:   Allow at least forty-eight hours of recovery time between working a particular muscle group. For best results, wait at least five days before repeating a particular exercise. Working the same muscle group every day can make it weaker rather than stronger.

THE SEVEN BEST GH-RELEASING BARBELL EXERCISES

These exercises are the heart of a growth hormone stimulating program. As the graph on page 257 shows, lifting at 70 percent of maximum capacity triples the output of GH, while lifting at 85 percent of max quadruples it! In this 1991 study done by Dr. William Kraemer and his associates at Pennsylvania State University, both men and women had the highest output of growth hormone when they did sets of eight to ten repetitions with one minute rest in between. Make sure your weight lifting routine includes the following exercises:

1: *Standing curl* (biceps)
Use a slightly wider than shoulder-width grip with your elbows locked to your sides. Curl bar until your biceps are at peak contraction. To work outer biceps use a close grip, for inner biceps. use a wide grip. Three sets of five to nine reps.

2: *Bench press* (chest)
Keep your body flat on the bench. Use a slightly wider than shoulder-width grip. Inhale as you lower bar to your lower chest. Keep elbows close to sides as you move. Exhale as you press the weight up. Three sets of eight to ten reps.

3: *Overhead press* (shoulders)
Using a slightly wider than shoulder-width grip, lower the bar behind your neck in a controlled manner to the upper traps (the trapezius shoulder muscles). You can also do this exercise lowering bar in the front to the upper-clavicle area. Keep your elbows pointed down, not back. Three sets of eight to twelve reps.

4: *Deadlift* (back, legs)
This can be done with one of two shoulder-width grips: an over/underhand grip in which one hand is over the bar and the other hand under, or a double overhand grip.
Place feet about shoulder-width apart, with arms locked and fully extended. Squat down until your thighs are about parallel to the floor. Your back should be tight, your head up and looking forward, your chest slightly forward and over the bar. Straighten your legs and raise the weight off the floor, keeping the barbell close to your legs throughout the movement. As you bring your legs closer to full extension, straighten your back to an upright position. Three sets of four to six reps.

5: *Squat* (thighs and buttocks)
Stand under a squat rack and rest the bar at a comfortable spot on your traps. Feet should be about shoulder-width apart and

slightly turned out. With your lower back arched slightly, head up and looking forward, squat down until your thighs are about parallel to the floor. Keep your knees over your big toes as you move. Raise your heels slightly to increase stability if needed. Gradually increase to three sets of six to ten reps.

If you want to add to the muscles of your outer thigh, keep your feet and legs together. If you want to hit the inner thigh, separate your legs about three feet apart with the feet turned outward. Remember, the knees stay in a line over the big toes as you move up and down.

6:   *Hack squat* (thighs and buttocks)

Raise your heels slightly and keep your upper body erect. Position the bar right up against your lower glutes (gluteus maximus), where the bottom of the buttocks meets the upper hamstring muscle in back of the thighs. Squat until your thighs are about parallel to the floor. Come back up, maintaining constant tension on your quads. Three sets of ten to fifteen reps.

7:   *Shrug* (shoulders)

Use a shoulder-width grip, keep arms locked, and shrug—don't roll—your shoulders straight up toward you ears. Three sets of five to seven reps.

The two power squats are especially good for growth hormone release. We also recommend leg presses on a weight machine, where you lie on your back and press up on the machine. These lower-extremity exercises in men recruit the large muscle groups around the groin, stimulating the male sex hormones, which in turn stimulate and are stimulated by growth hormone.

Your goal is to increase the resistance level to the point where you have exhausted your strength and you just can't do an extra rep. When you develop a better strength level and can handle more weight, you can try the next level listed below.

THE SCHWARZENEGGER EFFECT

Arnold Schwarzenegger knows all about growth hormone release. He did not need a lot of fancy equipment or advanced calculations to measure his VO2 max or a blood sample to determine lactate lev-els or GH stimulation. All he needed was to pump iron until he threw up. (He kept a barf bucket next to his barbells.) The rush of nausea and the vomiting that followed was the unmistakable sign of a growth hormone surge. Luckily (or maybe unluckily), you won't need to worry about the Schwarzenegger effect since only young body builders at high levels of effort and conditioning ever experience this. But if you want to go for the guts and glory, here's how to do it:

Do high intensity heavy lifting, three to five repetitions, two to five sets, at 70 to 85 percent of the maximum weight you can handle, three times a week. Once a week, do a 1RM, which is a one-time, all-out repeti-tion of the maximum weight you can lift.

Cautions

These are dangerous, potentially joint-damaging exercises but they are the best ones for growth hormone release. These exercises should only be done under the supervision of an athletic trainer.

MULTIANGULAR TRAINING

Phil Goglia is a nutritionist and bodybuilder guru to the muscle superstars and elite athletes. He has found that one of the best ways to achieve maximum GH levels is with multiangular training. The idea is to exercise just one body part, such as your arms, at many different angles, such as forty-five degrees, sixty degrees, and ninety degrees, and do it for almost an hour to full exhaustion. This is done using resistance machines and performing exercises at different angles. To ensure complete recovery of the muscle, that body part should not be exercised for a week. The rest and recovery time is extremely important, since overdoing it would *lower* the GH level.

GROWTH HORMONE EXERCISE RX

Here is the best way to get the highest peaks and the longest dura-tion of growth hormone secretion. If you are obese or a sedentary older person, start out with fifteen to twenty minutes of brisk walk-ing. As you get into better shape, increase to a twenty-minute race-walk or jogging and add ten minutes of weight lifting exercise. For those who are more fit, start with a thirty-minute jog-run and fifteen minutes of weight-training, increasing ultimately to forty minutes of running and twenty to thirty minutes of weight training. It requires a workout of a little more than an hour five times a week once you are in fit condition.

To get the fullest growth hormone enhancement effect, reverse the usual order of things. Begin with your warm-ups, progress to resistance training, and end with your aerobic exercise as follows:

1. Start with resistant exercises for about twenty to thirty minutes. This will get the spurts of GH going and tone and strengthen the body.
2. Then do thirty to forty minutes of aerobics that will maintain a high level of growth hormone for several hours afterward. At the same time, you get the cardiovascular benefit and augment the burning of fat.

WHEN SHOULD YOU EXERCISE?

For maximum weight-loss, muscle-gain effect, do what the bodybuilders do. In the morning when you wake up, drink water, preferably, or tea or coffee. Take some carbohydrate fluid for energy. Then do the GH workout outlined above. In this way you burn off yesterday's calories. And you get your thermogenic and metabolic rate higher for the day. When you finish your workout, eat breakfast, following a high protein, low fats, moderate carbohydrate menu for optimal body contour. To help replace protein lost during exercise, you can supplement with branched-chain amino acids at this time. (see "Metabolic Enhancers" below).

SCHEDULING YOUR WORKOUTS

The optimal weekly schedule is: Exercise for three days, rest one day, exercise for two days, rest one day. Alternate resistance exercise so that you don't work the same body parts in a row. For example, shoulders and arms one day, chest and back the next day, legs and abs the day after. Then rest one day and start the weight-lifting routine again.

METABOLIC ENHANCERS

Substances that your body makes naturally can improve your performance at high activity levels. As nutritional supplements, they help speed recovery of muscles and enhance cellular metabolism. The following eight substances are recommended by sports nutritionists, have research to back up their use, and are all available from the health food store.

L-Carnitine

This is a vitaminlike nutrient related to the vitamin B-complex, which is essential for energy production and metabolism of fat. It is crucial for anyone doing aerobic endurance exercise because it is involved in the oxidation of fatty acids, which supply most of the energy for endurance training. It is also of benefit in long weight-training sessions because it helps the body use branched-chain amino acids (see below). Carnitine has been shown to help reduce body fat in athletes and bodybuilders. Research in Japan and Europe also shows that it helps lower high triglycerides.

Although carnitine is a nonessential amino acid that can be synthesized from other amino acids in the body, it requires the presence of lysine, methionine, vitamin C, niacin, vitamin B–6, and reduced iron. Deficiencies can develop if the body does not make enough. The best sources of carnitine are meat and dairy products. Vegetarians who eat little dairy can get carnitine from fermented soybean products.

Take only L-carnitine as a supplement. D-carnitine or DL-carnitine, which is cheaper, has a negative effect on metabolism and has been shown to induce cardiac arrhythmias and muscle weakness. L-carnitine does not have these side effects and is completely safe at recommended dosages.

Dosage

1 to 2 grams a day for athletes to match their increased muscle mass.

Side effects

No known toxicity. If experiencing gastrointestinal intolerance due to salt content, cut back on dosage and take with meals.

Co-Enzyme Q10 (Ubiquinone)

CoQ10 is not only a metabolic enhancer, it is an anti-aging drug that has prolonged the mean life span of rats and mice. It is also an antioxidant, helps prevent heart disease, lowers blood pressure, improves the gums and reduces periodontal disease, enhances immune function, and stabilizes aging cell membranes.

It is important for exercise because it is a vital co-factor in the production of cellular energy. The body manufactures CoQ10 and it is found in every cell of the body. Deficiencies can occur, which can rob you of the energy you need for physical performance, especially in endurance activities. It also benefits exercisers by improving aerobic capacity, reducing fatigue, and helping to control the accumulation of fat and reducing body fat.

While CoQ10 is found in many foods, such as eggs as well as rice bran, wheat germ, fatty fish, organ meats, and peanuts, it is difficult to raise your levels solely through eating. The usual supplement is 30 milligrams, although a 1981 study showed that supplementation of 60 milligrams can improve exercise capacity by up to 12 percent after four to eight weeks of supplementation and training.

Dosage

Sixty milligrams for general muscle repair. One hundred milligrams a day for people with concerns about cardiovascular disease or for anti-aging effects.

Side effects

No known side effects or toxicity.

Chromium

This mineral is effective in increasing insulin sensitivity to blood glucose and overcoming the insulin resistance that affects most of us after age thirty-five. Chromium binds the insulin to the receptor sites on the cell membrane. Remember that insulin acts with growth hormone to build muscle protein. If chromium and glucose tolerance factor (GTF)—of which chromium is a part—is not present, insulin will not work properly and protein catabolism (breakdown) will continue unhampered. GTF also aids in turning carbohydrates into glucose, which is needed for energy. Exercise causes the depletion of minerals, which is why mineral supplementation is necessary. If you train vigorously, you are at risk for repeated losses of chromium. A 1984 report shows that 80 percent of the mineral is lost or removed in food processing. If you eat a diet high in simple sugars, you increase the loss of chromium by 10–300 percent. Brewer's yeast is an especially good food source along with whole grains, meat, orange and grape juice, broccoli, and thyme.

The best form of chromium supplementation is with niacin-bound chromium or chromium nicotinate. It is easily absorbed and has the greatest biological activity and safety. The usual dosage is 200 micrograms daily. Chromium nicotinate benefits include improving protein, fat, and carbohydrate metabolism; lowering cholesterol; increasing muscle mass; and generating energy.

Dosage

Keep under 1,000 micrograms a day. Tested dosage is 200 micrograms a day. Best taken as niacin-bound chromium or chromium nicotinate. Do not take simultaneously with calcium carbonate, which can impair absorption.

Side effects

Over 1,000 micrograms may interact (compete) with other minerals such as iron. Rare adverse effects include rashes, stomach ulcers, and poor kidney or liver function.

Caution

Diabetics should consult their doctors since chromium supplementation can affect insulin requirements.

Creatine

Creatine is an essential component of the high-energy creatine phosphate that stokes the energy battery in muscle and nerve cells—ATP. Creatine phosphate helps increase strength and speed in high-intensity, short-duration exercise. Supplements of creatine can increase the stores of creatine in the muscles and are easily absorbed into the muscle and nerve tissue. Creatine is popular among competitive athletes from Russia and Eastern Europe. It is used mostly for intensive physical training sessions.

Dosage

4 to 5 grams a day. Works better when taken after exercise.

Side effects

Weight gain typically seen is usually muscle weight gain from increased protein and water. If bloating or gastrointestinal cramping occurs, spread out dosage or take with carbohydrates such as sugar or simple starches.

Ginseng

This herb has been used for thousands of years in China, Japan, and other parts of Asia to promote general health and to treat a wide variety of illness. It is an adaptogen, which helps the body resist mental and physical stress. It normalizes body systems, balances the hormones, and boosts immunity.

Ginseng is especially good for exercisers because it greatly increases the capacity for both physical and intellectual work, with its effect on work capacity lasting up to thirty days after you've stopped taking it. Unlike caffeine, it is effective without producing a noticeable jolt. In combination with vitamins and minerals, it improves nerve function, has anti-inflammatory action, speeds wound healing, improves immune function, and increases synthesis of proteins and nucleic acids.

There are two types of ginseng—Asiatic ginseng (*Panax schinseng*) and American (*Panax quinquefolium*). Both are equally effective and safe.

Dosage

200 to 400 milligrams a day. Look for standardized extracts.

Side effects

No known toxicity, but some postmenopausal women may get estrogenic effects that can bring on the onset of the menstrual cycle. High doses can cause nervousness and insomnia.

Dibencozide (Coenzyme B–12)

Coenzyme B–12, as it is more commonly known, is directly or indirectly involved in all anabolic reactions. Weight lifting and endurance activities place a great anabolic strain on the body to synthesize protein and build new muscle tissue. Coenzyme B–12 along with proper diet, is used by serious bodybuilders to increase appetite and body weight. Its functions in the body include forming the myelin sheaths of nerves, and forming the bases, purine and pyrimidine, that are components of nucleic acids.

Dosage
1,000 micrograms a day to improve mood and feel less tired.

Side effects
Shown to be safe orally.

Gamma Oryzanol (Esterified Ferulic Acid)
Better known simply as ferulic acid, gamma oryzanol is found in rice germ, rice seed coat, and corn. It is a component of the lipids in these substances. In a 1990 double-blind study, ferulic acid supplements improved strength and increased body weight in weight lifters.

Ferulic acid improves physical performance and body function by enhancing blood circulation, decreasing workout fatigue, delaying muscle soreness, helping muscle buildup, increasing resistance to stress, and decreasing blood cholesterol. It also has antioxidant activity, which protects red blood cells, and it improves the ability of white blood cells to destroy foreign invaders.

Dosage
200 to 500 milligrams a day. Larger dosages have been shown to lower cholesterol.

Side effects
Shown to be extremely safe.

Branched-Chain Amino Acids
There are three branched-chain amino acids (BCAA), leucine, isoleucine, and valine. They get their name from the fact that they have interlocking methyl groups. The BCAAs are essential amino acids. Leucine deficiency impairs the function of the liver, thymus, adrenals, and gonads; isoleucine deficiency adversely affects nitrogen balance; valine deficiency throws off muscle coordination and makes one abnormally sensitive to sensory stimuli.

There is some research indicating that BCAAs may play an important role in protein synthesis and minimizing protein breakdown. Thirty percent of the protein in muscle cells is BCAA. Insulin transports BCAAs into the muscle cells, where they are used up rather quickly. If you are dieting, the BCAAs are the first to be expended either directly or by being converted into other amino acids. Even healthy people who are not dieting may need supplementation. First, exercise draws on amino acids, which need to be replaced. Second, the BCAAs don't hang around too long in the muscle cell, because they are easily converted to other amino acids but they can't be converted back.

If you are eating high levels of good-quality protein such as egg whites or whey protein, chances are you won't need supplements of BCAA. But they may be of help on days when your protein needs are not met. BCAAs come in capsules, powders, and synthetically designed food supplements like Amino Rx made by Biosyn Laboratories, which are easily absorbed into the bloodstream.

Dosage

2 grams a day for the average person for anti-aging purposes. Serious bodybuilders may take up to 30 grams of BCAAs.

Side effects

None with 2 grams. In higher doses can overload the kidneys. When taken on an empty stomach, it can cause cramping, bloating, and diarrhea.

HAVING IT ALL

A growth hormone enhancement program that combines GH-releasers, proper diet, and GH-stimulating exercises will allow you to shape your body like a master sculptor, chiseling away unwanted bulges, defining your muscles, and recreating the contours of youth. You will also be stronger and have more endurance than you've had in decades. Like Bob Delmonteque, you may find that you look and feel better now than you did in your twenties. Why not give your children or grandchildren a run for their money?

A Shot at Youth

About one-third of the aging population secretes little or no growth hormone. If you are part of this third, the chances are that the natural growth hormone agonists that we reviewed in Chapter 16 will not raise the levels of IGF–1 sufficiently to have age-reversing effects. In this chapter we will discuss growth hormone replacement with recombinant HGH injections, which should also be combined with the dietary measures and GH-releasing exercises to get the maximum muscle-building benefit. HGH is the product for which there is the most experimental and clinical experience. Although work is now being done with recombinant IGF–1 and GHRH (growth hormone–releasing hormone), which may prove even more effective than growth hormone itself, this is still very preliminary. And in Chapter 21, we'll tell you about the thrilling clinical studies now being performed on the secretagogues, which promise to transform the way in which all of us age—or don't age—by the turn of the century.

As we mentioned in Chapter 1, in August, 1996, the FDA approved HGH use for pituitary-deficient adults. Before this it could only be used to promote growth in GH-deficient children. The new indication is for SDS, or somatotrophin (growth hormone) deficiency syndrome, as a result of pituitary disease, hypothalamic disease, surgery, radiation therapy, or injury. In effect, the FDA approval covers the use of HGH for anti-aging purposes since low levels of growth or IGF-1 indicate a failure of the pituitary to release adequate amounts. In addition, the signs of SDS, such as decreased physical mobility, lower energy, higher risk of cardiovascular disease, are the same as those seen in aging adults with low GH levels. The new FDA indication means that the U.S. joins 12 other countries in which HGH replacement in adults has been approved, including Sweden, Denmark, France, Germany, Mexico and the United Kingdom. While Eli Lilly is the first drug company to receive approval from the FDA for the new indication, it is expected that the other manufacturers of HGH will soon follow suit.

GROWTH HORMONE MANUFACTURERS

Genentech and Lilly were able to rule the roost in marketing growth hormone for seven years in the United States under the Orphan Drug Act (see Chapter 2). But in 1992 for Genentech and 1994 for Lilly their exclusivity expired and other companies began scrambling to get into the act. Genentech, which commands about 75 percent of the $350 million growth hormone market, waged court battles to keep two would-be competitors—Novo Nordisk from Denmark and Bio-Technology General Corp (BTG) of Iselin, New Jersey—out of the United States by claiming patent infringement. Novo Nordisk's growth hormone has now successfully penetrated the U.S. market although BTG as of this writing had not. These are the five companies that now sell growth hormone (or most likely will do so in the near future) and their products:

1: Genentech, Inc., 460 Point San Bruno Boulevard, South San Francisco, CA 94080.

Genentech manufactures three growth hormone products—Protropin, the first recombinant HGH to be manufactured in 1985; Nutropin; and its latest drug, Nutropin AQ (aqueous). The last one is the only growth hormone available in liquid form for injection rather than as a freeze-dried powder that has to be first mixed with water. Nutropin also has a greater stability, lasting twenty-eight days rather than the fourteen days with Protropin. Genentech's HGH contains 192 amino acids, one amino acid more than the growth hormone produced by our bodies. This may explain why 30 percent of children taking it develop antibodies to the hormone, although the antibodies are not associated with adverse side effects and in general do not interfere with the body's response to the drug.

2:   Eli Lilly and Company, Lilly Corporate Center, Indianapolis, IN 46285.

Lilly was the first company to produce a recombinant DNA growth hormone product that has the identical 191–amino acid sequence as that produced by the human pituitary gland. The product, called Humatrope, has caused the formation of antibodies in only 2 percent of children using it.

3:   Pharmacia Inc., P.O. Box 16529, Columbus, OH 43216.

Pharmacia's product, Genotropin, was approved by the FDA in August 1955. Pharmacia and Upjohn, which merged last year, escaped court challenge because Pharmacia's Swedish affiliate, Pharmacia AB, licenses the growth hormone patents from Genentech. It features several sophisticated delivery systems, including the Genotropin Pen, which is has a cartridge for mixing the growth hormone powder and a small needle at the end for jabbing the skin.

4:   Novo Nordisk, 100 Overlook Center, Suite 200, Princeton, NJ 08450.

Norditropin went on the market in July 1995.

5:   Serano Laboratories, Inc., 100 Longwater Circle, Norwell, MA 02061.

At this writing, Serono has two NDAs (new drug applications) pending approval by the FDA for its growth hormone product, Serostim. One NDA is for growth hormone deficiency in short-statured children and the other is for AIDS wasting in adults. As an AIDS drug, Serostim is on the FDA's "fast track" review, which was set up to expedite the approval process for lifesaving drugs. Serano is a member company of the Swiss-based Ares-Serono Group, which has been involved in growth hormone research and development for more than twenty years. It uses mammalian cells to grow its growth hormone, thus getting around Genentech's patented method of making it in the bacterial cells of *E. coli.*

COST OF HGH

Growth hormone remains one of the most expensive drugs, costing anywhere from $800 a month to up to $18,000 a year for adult use. In its 1995 annual report, Genentech lists its sales as $219.4 million, and that's just for treatment of the uncommon condition of short stature due to growth hormone deficiency, which occurs in about one of every 5,000 children. The new indication for use in adult GH-deficient patients and the possible approval of HGH for such conditions as osteoporosis and obesity should push the potential market beyond that of any prescription drug in use today. As new drug companies rush into the fray and competition increases among them, the price should drop sharply.

Cheaper sources of the drug are available now from other countries, such as Mexico and Venezuela. According to Allan Ahlschier of San Antonio, Texas, "You can walk across the border from Texas into Mexico and get good, bioactive growth hormone in any drug store over the counter. Eli Lilly's product, Humatrope, sells for $43 for four IU [international units], and $130 for 13 IU." Depending on the dosage, a month's supply at these prices could run from about $150 to $400, or less than half what it costs in the United States. To save on costs, some U.S. growth hormone users are going to clinics in Mexico, such as the Renaissance Rejuvenation Centre in Cancun.

At least two companies are making HGH directly available to the consumer. The Swiss Rejuvenation Centre, Inc., which is a Bahamas corporation, is supplying growth hormone to patients in three-month supplies for their own personal use. Their price, which fluctuates, is about $25 per IU. For information on how to obtain this mail-order HGH, contact: Sonoma Diagnostic Inc., P.O. Box 1448, Delray Beach, FL 33447. Phone: (561) 241–0789 or (800) 635–3021, or fax (561) 995–8583.

Sportkoop, a Lithuanian-based company, produces HGH under the tradename Somategen-L. Their price is $15 per IU, which at a dosage level of four units a week comes to about $3,000 for a year, putting it within the reach of most people who are concerned about reversing the ravages of aging. This product is available from several overseas pharmacies (see Appendix for suppliers).

(Also see Appendix for names of manufacturers of growth hormone.)

DOSAGE OF GROWTH HORMONE

We suggest that you should first try raising your own endogenous levels using growth hormone releasers (see Chapter 16). These are safe, nontoxic supplements that will not interfere with your pituitary function, and they have the added benefits of being about one-tenth the cost and not requiring injection.

If you do not respond to these releasers and continue to have low IGF–1 levels and you decide to opt for growth hormone replacement, be aware that the therapeutic doses of growth hormone for age-related GH deficiency have yet to be determined. Most researchers agree that the dosages used by Daniel Rudman in his groundbreaking study were too high, causing unacceptable side effects such as carpal tunnel syndrome and gynecomastia. Rudman himself was planning to use lower doses in his subsequent studies. A press release issued by the Office of Public Affairs at the Medical College of Wisconsin in November 1993 notes, "The results of our clinical trials with HGH have revealed that the optimum hormone dose is only one quarter to one half as great as was previously believed. At the lower doses of HGH, a beneficial enlargement of muscles and shrinkage of fatty tissues is achieved without risk of adverse side effects" (carpal tunnel syndrome, enlargement of the male breast, diabetes, or fluid retention.) Other researchers, including the original groups in England, Denmark, and Sweden, went through the same learning process, starting out with amounts that were too high and resulted in side effects and finding out that they could achieve the same benefits with a lower dosage without adverse or uncomfortable side effects.

In this book we are concerned only with *replacement* doses, putting back into your body what nature has taken away. Going beyond replacement doses of GH or IGF–1 as some athletes, particularly bodybuilders, are doing is dangerous and foolhardy (see Chapter 20). The best approach to growth hormone replacement, as indeed to any hormone replacement, including that of insulin, is to start low and incrementally adjust the dose to the level that is suitable for each patient based on IGF–1 levels, clinical response, and side effects.

The good news for your pocketbook is that dosages needed may be far lower than what is generally used. A 1995 study by Hans de Boer and others at the Free University Hospital in Amsterdam came up with an optimal replacement dose in growth hormone–deficient adults. They based this on the resistance to a low electrical current passed through electrodes placed on the hand and foot. The technique, called bioimpedance analysis, is used to measure extracellular water (ECW). Low ECW is associated with high bioelectrical resistance. People who are deficient in growth hormone have low ECW and high bioelectrical resistance. With growth hormone replacement, rehydration occurs, the amount of extracellular water increases, and the resistance increases. However, if the dose is too high, the extra fluid volume causes many of the side effects related to growth hormone such as joint pain and carpal tunnel syndrome. Normal rehydration and ECW occurred with dosages of 1.10 IU. (An IU is 3 milligrams.) They recommend starting with an initial dose of .5 IU a day and raising it in monthly increments of .5 to 1 IU until a maintenance dose of approximately 1.10 IU a day is reached. This dosage is about 40 to 60 percent lower than that used in most previous studies, yet it accomplishes virtually the same results with no fluid retention problems.

Bengtsson believes that the dosage for older GH-deficient adults may be less than half of that, since the Dutch study was based on growth hormone–deficient men between the ages of twenty and forty. "The older you are," he says, "the more susceptible you are to growth hormone. A half unit a day should be sufficient for someone who is over age 60." Chein uses what he calls the high-frequency, low-dose method, which is one shot of HGH twice a day. About 90 percent of his patients use between 4 and 8 units a week, with no one exceeding 12 units a week. With this replacement dose, he says that he achieves IGF–1 levels that are equivalent to those of a twenty- to thirty-year-old. He has his patients inject themselves before going to sleep and again on awakening. They do this for six days, cycling off the hormone on the seventh day to avoid turning off the pituitary gland. Ahlschier, who considers his dosages one of his "trade secrets," says that if you over 2 IU a day you are going to have noticeable side effects.

The prudent course with any hormone replacement is to start low and increase slowly, if needed. Based on the de Boer study, a good starting point is a half unit a day, increasing it to 1.0 unit after a month. Bengtsson and others treat on an everyday basis, because your body releases growth hormone every day, not three times a week. It is probably a good idea to go off the hormone one day a week to ensure that the body's own machinery for releasing growth hormone remains functional.

The experience of most physicians in treating either growth hormone–deficient adults in Europe or the elderly in this country would indicate that doses above 8 units a week are probably excessive for most people, although it is important to note that there can be considerable individual variation. Cut back if there is any uncomfortable fluid retention or excessive changes in body composition that would be on the acromegalic side.

HOW HIGH SHOULD YOU GO?

Vincent Giampapa believes that replacement should be based on the average growth hormone levels in people aged thirty to thirty-five. "We try to keep it within that range. There is some variability there, but we would not want to go to the 20-year-old level. We chose the early to mid–30s, because that is a time when development is over but the disease processes and age-related diseases have not yet made their appearance. It is also a time where there is the lowest incidence of cancer. After that you start seeing the curve for disease processes and age-related diseases going up. We feel that this is a safe level of replacement for many of the hormone levels, because if those levels were associated with cancer and other disease processes, you would see them showing up in the general population."

It is best not to exaggerate the body changes beyond the levels noted here so that one crosses over the line into acromegaly. In the quest for youth, aging people can become like overcompetitive athletes or anorexic teenagers, losing perspective as they push their bodies into higher and higher goals that increasingly depart from reality. Our aim in this book is to promote health, not wreck it.

At replacement doses back to the levels of a normal thirty- to forty-year-old, a person who is deficient in growth hormone should experience all the benefits talked about in the first half of this book, an age-reversal equivalent to turning the clock back one to two decades. Is this the limit? No one knows, since growth hormone replacement even in adults with pituitary disease goes back only about seven years. Many of the benefits continue unabated, and according to Bengtsson, who has the longest experience with GH, his patients keep doing better and better. Khansari and Gustad showed that growth hormone significantly extended the average life span of mice (see Chapter 5), and the work of William Sonntag and colleagues in caloric-restricted animals points to growth hormone as a major cause of the disease prevention and dramatic extension of the maximum life span in these animals. It could be that growth hormone replacement will have youth-restoring and life span–prolonging benefits beyond the hopes and dreams of even its most ardent advocates.

THE POTENCY OF A POWDER

But for most of us, expensive and inconvenient shots are not necessary. Karen, a fifty-year-old mother of two and the head of her own small PR firm, decided to try growth hormone releasers rather than spend $1,000 a month on GH. She began taking 2,000 milligrams of glutamine and adding a weight lifting program to her daily five-mile run. Within a few days, she says, "I felt newly energized. I'm not one of those people who enjoy exercising, but suddenly I was looking forward to it. About a month later, people were remarking on how well I looked. My jeans, which were getting tight around the waist, suddenly had room to spare. And I was sleeping better than I had in years." Now, six months later, she has taken two inches off her waist and her friends are laughingly accusing her of having had a face-lift when she went to Mexico for a week. As for Karen, she is still shaking her in disbelief that her nighttime cocktail of a teaspoon of powder dissolved in water could work such wonders. "I call it the potion of youth," she says.

Minimizing Side Effects

Nothing in life is risk-free. When you ride the ocean waves, there is always the chance you may be eaten by a shark.

Even staying at home is no guarantee that you won't run into trouble, since that's where most accidents occur. What is important is knowing what the risks are and the reasonable precautions you should take.

But first let's look at some of the common concerns about growth hormone.

1:   *Growth hormone can cause acromegaly.*

André the Giant (née Roussimov), the man who played the giant in the film *The Princess Bride*, was an acromegalic casting a huge shadow at seven feet tall, weighing well over 400 pounds, with massive hands and feet. He was a wrestler who until a back injury in his late thirties could perform superhuman feats, lifting small cars. In an article in *Muscle and Fitness*, Terry Todd, Ph.D., describes his friend André asking why God would give a man something that made him so strong but would also shorten his life. André died shortly after at the age of forty-six.

André was born with a pituitary defect that poured abnormal amounts of growth hormone into his system. But athletes risk the same fate by literally pumping themselves up with growth hormone injections. Acromegalics are the inverse of adults with growth hormone deficiency, with decreased body fat, increased musculature, and increased total body water. But like the untreated GH-deficient patient they have a higher incidence of cardiac problems and premature death from heart disease than the normal population. They also have a higher risk of diabetes and cancer.

Acromegaly is not a side effect of growth hormone use. It is a direct effect of deliberate and continuous overdosing, doing everything the hormone is supposed to do but to great excess. While there are no studies that document increased death among athletes and bodybuilders due to abuse of HGH, physicians and researchers who work with athletes have seen the unmistakable signs of growth hormone overdosing in the bulbous noses and bulging foreheads of jocks and weight lifters. "These people are not taking physiological levels," says Dr. Robert Goldman, president of the National Academy of Sports Medicine and Special Advisor to the President's Council on Physical Fitness & Sports. "They are taking superhuman abusive levels that are ten to 100 times physiologic replacement levels. I've seen people die from this. One of them was a 44-year-old athlete who died of a massive heart attack. Growth hormone can be very exciting in the future in the senior population, but it is not indicated in young, hormonally balanced individuals prior to age 40."

But why would an athlete risk disease and death by using steroids and now growth hormone? In his book *Death in the Locker Room*, Goldman gave the answer. The question was put to 198 world-class athletes: If you could take a pill that would make you a winner at everything for five years but at the end of that time you would die, would you take it? Fifty-two percent said yes!

2: *Growth hormone is not an anabolic steroid but has similar effects.*

Anabolic steroids are synthetic products that act in the body like male sex hormones; they do not exist in nature. Growth hormone is not a steroid. It is a naturally occurring substance made by the pituitary gland, necessary for growth in childhood and an essential part of maintaining normal body composition and function throughout life. But growth hormone and anabolic steroids have the same effect on body composition, decreasing fat and increasing muscle mass. It is not surprising that growth hormone has become the drug of choice among athletes who are always trying to push the envelope of muscle mass and strength. That was why they abused anabolic steroids. But when testing began for anabolic steroids in competition, they had to turn to a compound that would have similar effects but would not be detectable. Growth hormone was the perfect answer. It cannot be detected in urine, and invasive blood testing is not permitted in competition. We feel that the use of growth hormone in young bodybuilders and athletes who have normal levels of GH is abusive, unethical, and dangerous. And it may be ineffective. In a recent study of experienced weight lifters who attempted to bulk up even further, short-term treatment with either growth hormone or IGF–1 did not increase the rate of muscle protein synthesis or decrease the rate of protein breakdown—changes in metabolism that promote muscle growth.

3:   *Growth hormone can cause edema and carpal tunnel syndrome.*
Carpal tunnel syndrome and joint aches and pains are all due to the water-retention effects of growth hormone. These are side effects and they have occurred in the past and continue to occur when physicians or researchers try to push the dosage. Rudman found that the dosages used in his study, which are several times that which are now recommended, causes these side effects. The recent double-blind study by Maxine Papadakis and associates at the University of California in San Francisco, which followed up on Rudman's study (see Chapter 18) used the identical dosage and, not surprisingly, got many of the same adverse effects. These included lower-extremity edema and joint pain. Two growth hormone recipients (and two placebo recipients!) developed tender breasts, although none of the men in the study developed gynecomastia. There was no incidence of carpal tunnel syndrome, and all the side effects disappeared when the dosage was lowered by 25 to 50 percent.

Some groups have reported no side effects at all, while others have reported an incidence of side effects that ranged from 5 percent to about 70 percent. The side effects appear to disappear when low dosages are used and the dosages are tailored to the individual patients. In one study of twenty-two GH-deficient patients, who were given about 0.5 IU per kilogram of body weight per week, only one person developed edema and there were no other adverse effects. In another study four patients on the same dosage developed edema, which went down when the researchers lowered the dosage to .25 IU per kilogram of body weight. The study by the Dutch team headed by Hans de Boer (see Chapter 19) revealed that the optimum dosages was about 1.10 IU a day. They recommend starting with an initial dose of .5 IU a day and raising it in monthly increments of .5 to 1 IU until a maintenance dose of approximately 1.10 IU a day is reached to prevent overhydration. This dosage is about 40 to 60 percent lower than that used in most previous studies. Yet the effects at these lower dosages were comparable to those of the highest dosage. The take-home message is that everyone is different. By starting low and slowly building the dosage in small incremental steps until a maintenance dose is reached, the body has the time it needs to adjust to the added hormone.

4: *Growth hormone and cancer risk.*

The possibility of cancer is a serious concern for researchers and physicians involved with hormone replacement therapy. The use of unopposed estrogen (without progesterone) in postmenopausal women led to an increase in endometrial cancer. Some studies have shown a rise in breast cancer incidence with estrogen replacement, while other studies have disputed this (see Chapter 15). Testosterone replacement has led to increased risk of prostate cancer. Now there is concern that growth hormone replacement could trigger cancer cells to divide more rapidly and promote the growth of a tumor.

Some experimental evidence supports this idea. Both HGH and IGF–1 stimulated the growth of leukemia and lymphoma cells in culture. Large doses of growth hormone also induced cancer in laboratory animals and promoted tumor development in rats with chemically induced bladder cancer. In human beings, patients with underproducing pituitary glands who had conventional hormone replacement therapy *without* growth hormone had a lower mortality rate from cancer than the general population, while patients with acromegaly have an increased risk of cancer. In addition, there appeared to be a slight increase in leukemia in Japanese children treated with HGH, but when children who had increased risk factors for the disease, such as previous leukemia, were accounted for, there was no difference in the rate compared with the normal population. And GH therapy did not increase the recurrence rate of tumors in children who had become deficient in growth hormone when they were irradiated for brain tumors, leukemia, or other cancer.

But there are reasons to believe that growth hormone replacement might actually *prevent* cancer. Chein and Terry have seen no cases of cancer among their group of more than 800 patients, although these people are in an age group where cancer rates are expected to rise. Nor has there been a higher than expected rate of occurrence of cancer among European adult patients treated for GH deficiency. And in the one case study reported by Chein and Terry, multihormone replacement along with growth hormone resulted in a *normalizing* of the PSA levels of a patient with prostate cancer (see Chapter 3).

Some of the strongest evidence of a possible protective factor comes from rats who are fed about 40 percent less than the control rats. Ordinarily growth hormone release declines in old rats the same way it does in humans. But the old diet-restricted rats had growth hormone levels that were the same as those of young rats that were normally fed. And, according to the work of William Sonntag and his associates at the Bowman Gray School of Medicine of Wake Forest University in Winston-Salem, North Carolina, the increased growth hormone may account at least in part for why the diet-restricted animals live far longer, have delayed aging, and develop less cancer than rats allowed to eat all they want. And the life span study of David Khansari and Thomas Gustad at North Dakota State University in Fargo found that low doses of growth hormone in old mice increased their immune function and significantly prolonged their life expectancy (see Chapter 5).

In animal and human studies, there is a wealth of data showing that growth hormone revives and rejuvenates the immune system, including the cells that recognize and fight cancer cells. And remember that growth hormone levels are highest just before puberty, when the incidence of cancer is at its lowest. The rate of cancer rises with age at the same time the amount of growth hormone wanes.

But there are two other points to take into consideration when contemplating growth hormone replacement. First, the vast majority of men and women die of heart disease, not cancer. As we discussed in Chapter 8, growth hormone replacement in people who are deficient has been shown to strengthen the heart muscle and function and decrease the risk factors of heart disease. Second, cancers take years to decades to develop from a single mutated cell into a full-blown tumor. "It is still an unanswered question whether growth hormone increases the possibility of cancer," says Terry. "But my feeling is that if you are an older individual and you get 15 to 30 years of living in a more healthful way, you've gained more than the fear of a tumor."

REDUCING ADVERSE SIDE EFFECTS

By carefully following the advice in this book, you should be able to avoid most, if not all, the problems associated with growth hormone replacement therapy. This means choosing a responsible doctor, doing a good medical workup before starting therapy, and having regular follow-ups. Here are some tips gleaned from the largest study of patients on growth hormone replacement therapy.

1: *Replace GH in a natural way.* This means using physiologic, not pharmacologic, doses on a daily basis. Start with low doses and adjust accordingly, if side effects develop. Chein recommends what he calls the high-frequency, low-dose method. Ninety percent of his patients get between 4 and 8 IU a week, which they divide into two shots a day, one at bedtime to enhance the GH pulses during sleep and one when they wake up. They do this for six days and take the seventh day off to ensure that they do not turn off their natural GH production from the pituitary. No one takes more than 12 units per week. The result is that in his study with Cass Terry of more than 800 patients on growth hormone replacement therapy, there were almost no side effects.

Medical centers that do clinical studies run into trouble, according to Chein, because many of them are using *supra*physiological doses, that is, amounts far beyond that which is produced by the body. They also give the shots three times a week rather than every day. "It's called design practicality," he says, "meaning that there is no patient on this planet willing to come to a hospital medical center twice a day to get injections. But the pituitary gland does not secrete growth hormone on Monday, Wednesday, and Friday. It changes all the time in the body. The patients in the medical center studies have all been given growth hormone in an unnatural way and that is why they have side effects."

2: *Keep IGF–1 at a normal level for a thirty- to forty-year-old.* Most of the patients in Chein's study were in the high 200s or low 300s of IGF–1 units, which is about the mean for people in their thirties. At that concentration, they got all the wonderful benefits, such as increased muscle mass, strength, energy, endurance, and sense of well-being that are the true legacy of growth hormone replacement. To push beyond 350 is to invite the problems associated with growth hormone excess and acromegaly.

3: *Replace all the hormones that decline with age in a balanced way.* Growth hormone does not act in a vacuum in the body. It is a master hormone affecting the reproductive hormones and DHEA and in turn is affected by them as well as by melatonin. When we are young all these hormones work in harmony in the body. Bringing back all the hormone levels that decline with age, says Chein, restores the equilibrium that preserves health and prevents disease. For instance, he points out that when progesterone is given along with estrogen to postmenopausal women, the increased risk of endometrial cancer completely disappears. The same thing is true, he says, when DHEA, an anti-cancer hormone, is given along with testosterone. The addition of melatonin in people over age fifty may also lower cancer risk since the hormone is one of the most potent antioxidants ever discovered and it has been shown to have anti-cancer effects (See Chapter 15 for a discussion of all these hormones.)

SIDE EFFECTS FROM GH-RELEASERS

If you are using nutrient-based growth hormone releasers at the suggested doses, you should not encounter any adverse side effects. However, even the most innocuous substance, such as wheat or nuts, can cause a dangerous reaction in a vulnerable individual. And anything strong enough to produce a desired effect has the potential for producing an undesirable one. For this reason, it is desirable that you work with a physician.

If you wish to experiment with prescription drugs where there are known side effects, it is imperative that you see a doctor both to obtain the prescription and to monitor the effects. Again, the rule is to individualize the dosage by starting at the low end and increasing slowly, cutting back if problems arise. Know the risks and err on the side of caution.

Growth Hormone: The Next Generation

Growth hormone and GH-releasers are just the beginning. Two other hormones promise to have an equal, if not greater, benefit for the treatment of aging and age-related diseases: GHRH and IGF–1. These might be called the before and after hormones since GHRH precedes the release of growth hormone and IGF–1 follows it. Also under development are specific growth factors that are being used in the healing of everything from wounds to neurodegenerative diseases. Finally there are the true wonder drugs of the future, the secretagogues—powerful, cheap, oral drugs that promise to release growth hormone better even than injections of GH. They will put age-reversal within reach of every one of us. The first of these products should be on the market by the time this book is your hands.

GROWTH HORMONE–RELEASING HORMONE (GHRH)

This is the spigot that turns on growth hormone in the body, so it stands to reason that it may be the most natural method for inducing it in people whose levels of GH are declining. But, for the moment, it is used mostly for research purposes. Synthetic GHRH is made by Serano Laboratories in Nowell, Massachusetts, but it has not been approved as a drug in the United States. One Ph.D. researcher and CEO of his own company, who wishes not to be identified, injected himself for the six months on a twice-weekly basis.

"It is remarkable in its ability to stimulate your own growth hormone," he says. "You can see the change in musculature in adipose tissue within three to five days. The muscle definition increases very dramatically and the fatty tissue is reduced. The fat just burns away. It gave me increased energy and strength in the gym. It's scary and amazing. It makes you feel like a teenager. I'm forty-six but I'm probably the youngest looking middle-aged businessman in the world."

When he stopped using GHRH for four months, he still retained about 70 percent of the benefits. He is now going to start using it again. "Some people respond better than others. Sometimes it doesn't work if you don't have good growth hormone production, but my body responds very dramatically even if I use other stimulants like glutamine or GHB, which is a very good hormone inducer. But this stuff is probably as potent as anything there is."

THE UMBRELLA EFFECT

Think of GHRH as an umbrella with growth hormone under it, and below that insulin growth factors 1 and 2, and under that the myriad growth factors, which carry out the specific actions on the cells and tissues of the body, says Pierre Savard, Ph.D., associate professor of the faculty of medicine at Université Laval in Quebec, Canada. "The ideal therapy would be to give GHRH," he says, "which would have an umbrella effect on the stimulation all the way down the line. It favors the release of growth hormone which favors the release of growth factors at the level of the tissues."

Growth hormone–releasing hormone declines with age the same way that growth hormone and IGF–1 do. But several studies show that giving GHRH not only stimulates GH and IGF–1 but works as well in old men (and presumably women) as it does in young men.

GHRH RESTORES GH LEVELS

Emiliano Corpas, M.D., and his colleagues at the Gerontology Research Center, Francis Scott Key Medical Center, and Johns Hopkins University School of Medicine, all in Baltimore, Maryland, found that twice-daily subcutaneous injections of GHRH restored the levels of growth hormone and IGF–1 levels in older men to that of men three decades younger. The study involved nine men between the ages of twenty-two and thirty-three and ten men between sixty and seventy-eight. All the older men had reduced levels of GH and IGF–1 to start. The study was set up so that the older men were given injections of either low dose GHRH (.5 milligram) or high dose (1 milligram) twice daily for fourteen days, then no treatment for another fourteen days. At that point those who had gotten low doses received high doses and vice versa for another two-week period.

At the end of the study period, the men who received the high dose of GHRH had levels of growth hormone peaks and IGF–1 that were comparable to those of the younger groups. Six of the men on high doses had increases of more than 21 percent in their IGF–1 level. None of the subjects reported any problems or side effects with GHRH and there were no adverse changes in function, such as increased blood glucose or blood pressure. The researchers conclude that "short term subcutaneous administration of GHRH to healthy old men reverses age-related decreases in GH and IGF–1, suggesting that prolonged treatment could improve age-related alterations in body composition."

In a second study the Baltimore researchers used an infusion pump planted under the skin to deliver two weeks of either low-dose or high-dose GHRH. Both high and low doses of the GH-releasing hormone restored the subnormal growth hormone secretion and IGF–1 levels, particularly during the day. This type of treatment, they say, might "produce a physiological pattern of GH release, be more convenient than daily injections, and offer potential therapeutic advantages in the long-term treatment of older persons with subnormal GH and IGF–1 levels."

Another group of researchers found that doses of only 80 micrograms of GHRH given intravenously were enough to raise growth hormone levels significantly in a group of young healthy men without causing unwanted side effects. There was no difference in GH response above that dosage.

Right now GHRH is being used almost entirely for research purposes. It is available in synthetic form from Serano Laboratories and Hoffman-La Roche. And it is sold in pharmacies across the border in Mexico. It has been tested on pigs, where it increased the muscle weight by 16 percent and decreased fat weight by 25 percent. It also increased skin and bone weights by 29 percent and 19 percent, respectively. While the researchers in this case were interested in producing leaner pork, presumably what works for bacon should work for us as well. More studies, including long-term treatment, have to be carried out in humans before the hormone can be sold for therapeutic purposes. But the potential is great for GHRH because it more closely mimics the way growth hormone is released in the body, and the studies of the Baltimore group show that it can actually reverse the loss of growth hormone that occurs with age.

INSULINLIKE GROWTH FACTOR-1

This is the other end of the growth hormone chain, the downstream player that actually exerts most of the effects we associate with GH. IGF–1 is causing a great deal of excitement among two groups, researchers who are exploring its vast potential and bodybuilders who are already using it and claiming eyepopping gains in muscle.

MORE POTENT THAN GH

As we have said a number of times in this book, growth hormone exerts most of its effects through IGF–1. Therefore, it is not surprising that IGF–1 injections will do for you what GH does—and then some, according to its proponents. It increases lean body mass, reduces fat, builds bone, muscle, and nerves. By taking it directly, you bypass the pituitary gland, which may be "burnt out" with aging.

It appears to be even more potent than growth hormone in its anti-aging action. According to Keith Kelley, Ph.D., who did the work showing that growth hormone reversed the shrinking of the thymus (see Chapter 7), when he does his experiments on cells in culture, only IGF–1—and not GH—works. But both IGF–1 and GH work in the living animal. "I know that both GH and IGF–1 are substantially elevated in the old animals treated with growth hormone," he says, "but my prediction is that the main player is going to be IGF–1."

IGF–1 IMPROVES GLUCOSE METABOLISM

As its name indicates IGF–1, or insulinlike growth factor-1, has similar properties to insulin, and it has improved blood sugar profiles in type 2 diabetic patients. High doses of growth hormone have been shown to increase insulin resistance, but IGF–1 administration actually normalized the insulin resistance in a group of healthy volunteers.

In the latter study, Nelly Mauras and Bernard Beaufrere of the Nemours Children's Clinic in Jacksonville, Florida, were looking at several different things: the effect of IGF–1 on protein metabolism; its ability to stop the protein-wasting caused by glucocorticosteroid drugs like prednisone, and its effect on insulin and glucose metabolism. They divided the volunteers into three groups who got one of the following: IGF–1 alone, IGF–1 plus prednisone, and prednisone alone. The study found that IGF–1 at 100 micrograms per kilogram of body weight given twice daily enhanced the body's protein metabolism in the same way as growth hormone. Like growth hormone, it markedly decreased the protein breakdown in the volunteers who were taking prednisone. But whereas growth hormone in an earlier study caused carbohydrate intolerance and insulin resistance when given in combination with prednisone, IGF–1 did not cause these diabetes-like effects. Instead, those subjects who received IGF–1 along with prednisone had normal glucose metabolism. This was remarkable, say the researchers, in light of the fact that glucocorticoids are known to suppress circulating insulin and decrease insulin sensitivity. As a result of this and previous studies, the researchers believe that IGF–1 offers promise in the treatment of protein catabolic states, such as patients who require IV feedings after surgery.

IGF–1 REGENERATES NERVES

One of the most exciting potential uses of IGF–1 is the repair of peripheral nerve tissue that has been damaged by injury or illness. If a nerve is torn in the arm or leg, it means that the connection to the muscle may be impaired, and as result there is loss of movement and the muscle atrophies. While the peripheral nerves can regenerate to some extent, severe tears of more than a few millimeters may result in permanent injury. But IGF–1 has repaired and reconnected severed nerve endings of up to a distance of 6 millimeters, a feat previously unheard of.

Swedish scientist Hans-Arne Hansson of the Institute of Neurobiology at the University of Göteborg found that IGF–1 in combination with other growth factors such as PDGF (platelet derived growth factor) could stimulate even more dramatic regeneration. IGF–1 by itself and in combination with other growth factors is "likely to be of importance in promoting healing and repair processes in clinical practice within a few years," he writes.

In studies of cells in culture and in animals, IGF–1 has been shown to have remarkable effects on the spinal cord motor neurons. It increased motor neuron activity in spinal cord cultures by 150 to 270 percent. And it significantly decreased programmed cell death in developing chick embryos. In animal studies, it enhanced the sprouting of axons of the spinal cord motor neurons. And it increased intramuscular nerve sprouting a whopping tenfold when it was given to normal *adult* rats. In fact, according to a group of researchers at Cephalon, Inc., in West Chester, Pennsylvania, IGF–1 may be the "long-sought endogenous motor neuron sprouting factor."

The implications of this work for helping people is nothing short of mind-boggling. If IGF–1 can regenerate spinal cord motor neurons, it may be useful in treating amyotrophic lateral sclerosis (ALS), a devastating disease in which the loss of spinal cord and cortical motor neurons results in complete paralysis and death. It may also be useful for peripheral neuropathies, such as Charcot-Marie-Tooth syndrome. (The effect of growth hormone replacement in raising IGF–1 levels may account for the fifty-nine-year-old rancher's stunning remission from this disease as reported in Chapter 6.) And it may allow more aggressive chemotherapy of certain cancers, since drugs like vincristine and cisplatin can cause peripheral neuropathies at higher dosages.

THE BODYBUILDERS' DREAM DRUG

Clinical studies of IGF–1 are just now getting under way. John Wittig, M.D., of UCLA has been using it to prevent AIDS wasting in HIV-infected patients. But IGF–1 is where HGH was a few years ago. Physicians are still trying to work out dosages, control side effects, and find out how often it should be cycled, that is, when people should go on and off the drug.

But one group has already jumped on the IGF–1 bandwagon—the bodybuilders. A number of world-class bodybuilders are using IGF–1 and reporting massive muscle magnification of up to twenty pounds. An article in *Muscle Media 2000* trumpets IGF–1 as "Possibly the Most Potent Bodybuilding Drug Ever!" According to author T. C. Luoma, "IGF-1 is out there on the streets of America right now; it's being sold out of trunks of cars in Venice and brown paper packages containing it are being discreetly handed out at Southern California gyms." While there are no controlled studies supporting the musclemen's claims, the anecdotal evidence is building up like, well, biceps with a barbell. Says the article, "Bodybuilders are claiming that they're experiencing drops of 5% body fat in a *month* [emphasis added], while increases in lean body mass and strength are 'incredible.' Statements like, 'It's the most wonderful stuff in the world,' and 'I couldn't believe it man' are the norm."

But Mauro Di Pasquale, M.D., an expert in performance-enhancing compounds, is skeptical. "Studies are now being conducted on AIDS patients using up to a hundred times the amount of IGF-1 that bodybuilders are using and seeing relatively little anabolic effects," he says. "It makes you wonder about the so-called dramatic result some bodybuilders attribute to their use of IGF-1, especially at doses between 20 and 100 micrograms per day." He also points out that much of the IGF–1 on the market now is counterfeit and doesn't contain any active ingredient. "So any effect in these cases is not due to the IGF-1 but to other drugs used, such as growth hormone, insulin, and anabolic steroids as well as a number of other drugs and hormones such as thyroid, clenbuterol, DHEA, tamoxifen, diuretics and a large number of nutritional supplements."

It is not clear whether stacking the hormones, that is, using growth hormone and IGF–1 together, for example, will have a synergistic effect. In one study of healthy volunteers, there was no additive anabolic effect in those who took a combination of IGF–1 and GH, compared with those who took either GH or IGF–1 alone. And a study of athletes found that IGF–1 did nothing for increased muscle protein synthesis in experienced weight lifters, presumably because they already had lots of IGF–1 on board.

But there is at least a rationale for thinking that these two intimately related hormones might work better in tandem. There is a feedback control mechanism between the growth hormone in the pituitary and IGF–1 in the liver. The growth hormone stimulates the release of IGF–1, but when the levels of IGF–1 rise to a certain point in the circulation, it signals the shutdown of GH. But there is a lag time in all of this, which means that growth hormone levels increase at night and IGF–1 levels increase during the day. Bodybuilders hope that taking the two together will have a double-fisted effect on protein metabolism, decreasing protein breakdown and increasing protein synthesis. But although some studies have shown that this is the case in people who are calorically deprived, the study on healthy volunteers eating three squares a day showed no enhanced benefit from combining the hormones. It's possible, says Di Pasquale, that intense exercise may allow for a synergistic effect of GH and IGF–1, but that's still to be determined.

IGF–1 is such a new product that it seems that the FDA has yet to figure out what to do with it. It is neither available upon prescription nor a controlled substance. Reports indicate that human dosages vary widely from about 10 micrograms to 50 micrograms daily. According to Phil Micans of International Aging Systems in London, who stocks IGF–1, two forms of the hormone are available. One is a media cell culture grade, which has about 70 percent purity. The other is a receptor grade that has 95 percent purity. Although the latter is really the drug meant for human use, it is ten times as expensive as the cruder form. But Micans believes that IGF–1 will be the hormone of choice in a few years. It is *ten* times more potent than growth hormone, he says, so that you could reduce the dosages considerably to get the same anti-aging effects. And as more is known about it, more companies will produce it, inevitably pushing down the price.

"We don't sell a lot of IGF-1 right now," says Micans, "because people don't know what the heck it is. Even when they do know, they say, 'well how much should I take?' We say, 'when you find out, let us know.' It is really a little bit down the road."

THE GROWTH FACTOR ARMY

IGF–1 is only one of the body's many growth factors that are now being identified, isolated, and cloned using genetic engineering technology for use as drugs. As growth factor researcher Eric Dupont, Ph.D., says, "Growth hormone is the general and growth factors are the foot soldiers." Growth factors function like hormones, hooking onto receptors of cells and sending a biochemical signal across the cell's interior. But whereas hormones usually send long-distance messages, growth factors for the most part do local calls. The name "growth factor" is somewhat inaccurate since these proteins don't stimulate growth but may instead inhibit it or promote cell movement. To make matters even more confusing, different medical specialties call them by different names. For instance, proteins that actually enhance growth are known as growth factors to physiologists, cytokines to cell biologists, interleukins to immunologists, and colony-stimulating factors to hematologists.

These growth factors (GF) are strings of peptides, called polypeptides, with each GF possessing several biological functions. They have a complex, 3-D ribbon-like structure and it will take some time for biotech companies to untangle all these strands and see which section exerts what effect so that they can produce the specific desired result.

According to Savard, "The world growth factor market is a key area of growth for the biopharmaceutical industry. In 1992, it generated revenues exceeding $3 billion and was expanding at a rate of 80%." Many of these growth factors are already being sold or are now being tested as drugs to renew and rejuvenate cells and tissues that have been damaged by trauma, sickness, or aging. These include erythropoietin (EPO), which stimulates the proliferation of red blood cells, and is now being used in patients with anemia due to chemotherapy, AZT treatment for AIDS, or kidney failure; epidermal growth factor (EGF), which stimulates the development of epithelial tissue and is used to treat burns and ulcers, among other conditions; and fibroblast growth factor (FGF), which induces the formation of new blood vessels and is used to heal pressure sores and venous ulcers in skin graft donor sites.

Even more exciting is the development of neurotrophic factors, or nerve growth factors (NGFs), which can actually regenerate peripheral nerve and brain cells. There is growing evidence, says Savard, that a deficiency of these factors may cause the natural degeneration of the nervous system during aging. Neurotropic factors have also been linked to several neuromuscular diseases, such as amyotrophic lateral sclerosis (ALS), or Lou Gehrig's disease. Researchers have shown that the simultaneous administration of two nerve growth factors, BDNF (brain-derived nerve factor) and CNTF (ciliary neurotrophic factor) in Wobbler mice, the animal model for ALS, stopped the progression of this fatal, degenerative disease. Nerve growth factors also show promise in the treatment of the neurodegenerative diseases of aging, such as Parkinson's and Alzheimer's.

GROWTH HORMONE PILLS AND PATCHES

The Holy Grail of all this research is to find a small, cheap, safe pill that one could take every day to raise GH levels—a kind of anti-aging aspirin. This is the ultimate secretagogue, a fancy word for a growth hormone–releasing compound. And it would work in the body the same way that GHRH does, by telling the pituitary to release more of the growth hormone stores that are there, even in the elderly. Right now major pharmaceutical companies have joined the hunt for the perfect, oral, growth-hormone secretagogue, including Wyeth-Ayerst and Japanese company Kakan, Merck, Pharmacia Upjohn, Genentech, and Nova Nordisk.

Two are now well into clinical trials in this country. They are the brain children of two men who are good friends, Cyril Bowers, Ph.D., and Roy Smith, Ph.D., who are now developing secretagogue drugs for rival pharmaceutical companies hoping to produce what promises to be the biggest-selling drug in history.

BOWERS, THE FATHER OF GH SECRETAGOGUES

The story of the secretagogues begins with the story of Cyril Bowers, the acknowledged founder of the field of growth hormone–releasing peptides. Bowers is an indefatigable researcher with a puckish sense of humor and a touch of the poet. He started the ball rolling in the late 1970s, when he found that a modification of the Met-enkephalin peptide—one of the brain's morphine-like chemicals—acts on the pituitary to release growth hormone. This was a special discovery since such a small peptide of a known amino acid sequence—only five amino acids in length—meant that it could be made into a convenient package like a pill or patch to stimulate GH.

He and his colleagues at Tulane Medical School in New Orleans were able to utilize the structure of this compound by removing all of its opiate activity and increasing its ability to stimulate GH. Finally, after extensive research, they came up with a new version which had six amino acids. This was GHRP 6 and it worked well in animals and people. In fact, it worked so well, that "we thought we had invented growth hormone releasing hormone, GHRH," says Bowers.

Michael Thorner, M.D., chief of the division of endocrinology and metabolism at the University of Virginia Medical Center in Charlottesville, who played a crucial role in the discovery of GHRH, suggested to Bowers that he start testing GHRP 6 on people. GHRP 6 has now been tested on over a thousand people. It is very effective in stimulating growth hormone and is remarkably free of side effects. It accelerates growth of GH-deficient short-statured children, although not to the same degree that growth hormone does; it has a synergistic effect with GHRH, causing a far greater release of growth hormone than either one of these substances alone. When given with GHRH, it can overcome the block to GH release in severely overweight people (see Chapter 9). But what is most important to Bowers and Thorner is that GHRP 6 duplicates the way the growth hormone works in the body. When Thorner and his group gave GHRP 6 to normal young people, it not only enhanced their growth hormone levels but increased the pulsatile secretion—the bursts of GH.

Meanwhile Bowers now has a second-generation compound, GHRP 2, that is even more potent than GHRP 6. But preliminary trials in elderly volunteers have been somewhat disappointing. Raising the dosage did not cause the growth hormone levels to go higher as expected and when they pushed even harder, the levels of GH actually went down. However, it is possible, says Bowers, that giving GHRP over a longer time period will increase its effectiveness. "It's what I call the Kentucky Derby," he says. "At the top you have growth hormone, then IGF–1, then GHRH, and then GHRP." Then mixing his metaphors, he adds, "GHRP is David and growth hormone is the giant is at the top. It's hard to imagine when you have such a wonderful growth hormone that you could possibly compete with it. But possibly GHRP can compete and the place that it can is in the elderly."

To Bowers, GHRP is a kinder, gentler way to do what growth hormone does with brute force. "Everybody has been in such a hurry, trying to metamorphose elderly people in three months. If this is done, one will get into trouble with fluid retention and all the other side effects. With GHRP there should be fewer side effects and GHRP could bring back the normal pulsatile secretion of growth hormone but one must be patient and metamorphose these people over a period of time. I don't think that any company dealing with growth hormone could ignore GHRP. It is the most practical approach at the current time. It has one other beautiful feature and that is it may increase the normal pulsatile physiological secretion by its hypothalamic action. Thus it is easy to see why we have stars in our eyes."

HEXARELIN

But Bowers has some serious competition. One of these now being tested in clinical trials is Hexarelin, which is being developed by Pharmacia Upjohn. Like GHRP, Hexarelin (Hex) is a six–amino acid peptide that releases growth hormone. A group of scientists at the University of Turin in Italy recently showed that Hex significantly raised GH levels when it was given to normal young men and women in four different ways: intravenously, through subcutaneous injections, through the nose, and by mouth. This study showed that Hexarelin was more effective and longer lasting than GHRH. And it was dose-dependent, meaning the higher the dose, the greater the release of growth hormone. In fact, according to the researchers, Hex is even more potent than GHRP 6, although Bowers says that his studies indicate that the activities of both peptides are essentially the same.

MERCK'S ORAL GH-SECRETAGOGUE

Perhaps the greatest excitement in the field of secretagogues surrounds Merck's GH-releasing compounds. Inspired by Bowers's work at Tulane, Roy Smith, Ph.D., vice president of biochemistry and physiology at Merck and Co., in Rahway, New Jersey, decided to screen promising compounds from the collection at Merck that would mimic the activity of GHRP 6.

Smith and his team at Merck together with the medical chemi-cists came up with several secretagogues that appear to fill the bill. The most effective one is known as MK0677. It works the same way as Bowers's compounds. If this agent works as well over the long-term, Bowers wrote in an editorial on the Merck drug, "Serum GH and IGF–1 levels in older and younger subjects may become the same."

Like GHRP 6, MK0677 is a powerful, natural stimulator of growth hormone. But unlike GHRP 6, it does not have a typical peptide structure—and therein lies its advantage. Peptides, as we have mentioned, are small proteins. They are not well absorbed when you swallow them and your body treats them like any food, chopping them up into their individual amino acids during the digestive process. This makes it difficult to develop proteins as oral drugs. But a small molecule that contains no amino acids can easily be absorbed by the body without first being destroyed by digestion.

"This compound will induce the physiological pulsatile profile of growth hormone release," says Smith. "While GHRP 6 has to be infused intravenously constantly to get a pulsatile release, with this compound, you take a pill once a day. It amplifies the endogenous pulses so you rejuvenate the growth hormone axis. You can treat an elderly person and they will have the pulsatile growth hormone pattern of a young person."

The latest clinical trials in a group of men and women, ages sixty-four to eighty-one, showed that 25 milligrams of the drug taken once a day for two weeks raised the levels of IGF–1 from an average of 141 to an average of 219, with the highest levels reaching 251. And it appears to be well-tolerated by elderly people. The half-life of MK0677 is such, says Smith, that it hangs around most of the day. On the other hand, a peptide like GHRP 6 might have to be taken several times a day to be effective.

Another interesting thing about the Merck compound is that studies in animals show it is self-regulating. "The first time you take the drug, you get a relatively large outpouring of growth hor-mone," says Smith. "On day two, the response is much reduced, and by day three, the growth hormone levels and the IGF–1 levels have reached a plateau. It levels off in the normal range for a younger person." Could you give it every other day to get a higher response? "That depends," he says, "on whether you think that a big dose of growth hormone every other day is more important than maintaining a constant level of pulsatility. We focused on finding a compound that would reset the pulsatile levels, so we could rejuvenate the hypothalamic-pituitary axis by a once-a-day treatment. Intuitively, I think that would be the right approach for clinical use."

THE NEW UNDISCOVERED HORMONE

Bowers believes that the GHRP-GH secretagogues that his work produced could have a significance beyond the fact that they may be the basis for a new class of powerful anti-aging drugs. He thinks that they are telling us something unique about the brain itself. For a long time, scientists have puzzled over why growth hormone declines with age. The work with both growth hormone–releasing hormone, which releases GH, and somatostatin, which blocks GH, was not sufficient to explain the decline. Now Bowers may have the answer. GHRP 6, MK0677, Hexarelin, and all the other secretagogues bind to receptors that are different from the one that GHRH binds to. He believes that this indicates the presence of a new, undiscovered hormone that is a major regulator of growth hormone secretion. And when this substance declines in the body, it causes the release of growth hormone to go down.

In a stunning paper just published in *Science* magazine, the Merck group followed up on their earlier work showing that MKO677 and GHRP 6 acted on a new receptor by isolating and cloning the receptor that all the growth hormone–releasing peptides and all the nonpeptides bind to. "It appears to belong to a completely new family of receptors, says Smith. "MK0677 and the GHRP–6 compounds are active in chickens, rats, mice, sheep, cows, monkeys, and humans and experiments with the cloned receptor show it is expressed in the brain and the pituitary gland. The pathway is highly conserved in nature. And yet the natural hormone that stimulates this new receptor is unknown at this stage."

It may seem strange that a manmade compound could point the way to a completely new system in the brain, but that has happened before. Morphine was used for centuries and when scientists went looking for the receptors that morphine bound to, they found a series of opiatelike chemicals made in the brain, called endorphins and enkephalins, that bound to the same receptors in the brain that morphine did. "We think that just as there is an equivalent of morphine in the brain in the endorphins and enkephalins," says Thorner, "there is an equivalent peptide, probably a neurotransmitter produced in the brain that Dr. Bowers' compound mimics."

THE FIRST GHRP

No one knows exactly when the big drug companies will actually bring out an oral GH-releasing secretagogue. All the researchers in secretagogue research agree that interest is very high and competition great. But the FDA drug approval process, especially if the drug is not for a life-threatening disease, can take as much as eight to ten years. Chances are that the first prescription products will not make an appearance for at least a few years.

But you don't have to wait very long for a nonprescription growth hormone–releasing peptide that you can buy at your local health food store. This is an oral preparation that will be sold by only one company in the U.S. In clinical tests it has increased growth hormone levels higher than anything you can now buy without a prescription.

A new formula called AminoTropin-6 combines the nutrients GABA, arginine, lysine, and xanthinol nicotinate with a patented neuropeptide GH-releaser for a powerful synergistic effect. All the nutrients used in this preparation are well-documented growth hormone stimulants. Studies have shown that oral glutamine increases growth hormone by up to 15 percent. As we mentioned in Chapter 16, the combination of arginine and lysine bumps up GH release four times beyond that of arginine alone. A paper by Dr. A. Zsuzsnna in the journal *Endocrinology* reports that GABA increases GH secretion by about 5 percent. And xanthinol nicotinate is the most potent form of niacin for promoting GH release.

Dr. Ezio Ghigo and co-investigators in France found that the plasma levels of certain peptides determined the amount of growth hormone in the circulation. Scientists at several research centers in Western Europe have now shown that some of these peptides when taken orally have raised GH secretion by 30 percent. But now Aftab Ahmed, Ph.D., previously a faculty member at Northwestern University Medical School in Chicago, where he was instrumental in developing gene-derived therapies for disorders of the central nervous system, has gone one step further than the Europeans. He has patented his own GHRP, which he has included in this formulation, and linked it to glutamine and all the other nutrients listed above. Preliminary studies reveal that the GHRP-nutrient formula can raise growth hormone by up to 50 percent!

At a recent conference sponsored by Gero Vita International, a company that produces and distributes nutritional formulas for specific ailments, Ahmed, Hans Kugler, Ph.D., a research scientist and editor of *Preventive Medicine Update*, and Dr. N. Venkatesan of the University of California summarized the results of experimental and human data that AminoTropin-6 is capable of producing a 50 percent increase in GH levels. "This formula, relying on precise amino acid sequencing, appears to be truly synergistic," said one of the participating scientists.

Double-blind placebo controlled trials are now going on and should be completed by the time this book is published. According to the scientists at Gero Vita, "No side effects are expected from this new formula because AminoTropin-6 naturally stimulates the pituitary gland to release GH. The gland won't release more than the amount needed by the body. You don't have the same protection when you get injections of the synthetic growth hormone."

AminoTropin-6 may be a better and safer alternative to frequent injections of human growth hormone, concurred the scientists, and it costs a lot less—slightly over a dollar a day, compared with up to $40 a day of the synthetic hormone. "It appears to be an important breakthrough." (See Appendix for supplies.)

The new generation of growth hormone releasers is just the next step in the continuing journey toward a longer, healthier, more productive, more fulfilling life. The world of anti-aging medicine that awaits you is almost beyond imagining. But in the Epilogue, we will do our best to imagine it for you.

Epilogue

BRIDGES TO IMMORTALITY

At the start of the twentieth century, the average life expectancy was forty-eight years. Today the average life span in the U.S. is seventy-six years. In Japan, it is approaching eighty years. Just look around you at the people who are going strong in their eighties, nineties, even at age one hundred. In 1996 there were about 70,000 centenarians, and the number is expected to double by the year 2004.

Within the next thirty years, we can expect to see life spans of 120 to 130 years accompanied by good physical and mental health. The use of antioxidants, such as co-enzyme Q10, vitamin C, vitamin

Increase in Biological Knowledge

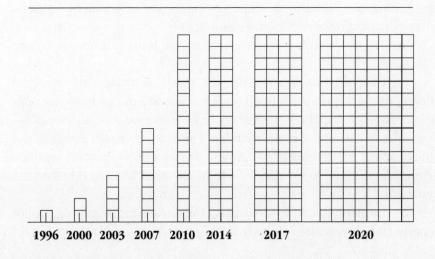

E, and the carotenoids, which actually get at a root cause of aging, the free-radicals that ravage the cells and cellular components, will help prevent and delay heart disease, cancer, and a host of other diseases and could add another ten years of functional life to the average life span.

The Growth Hormone Enhancement Program outlined in this book may help extend your health span by twenty-five to thirty years. This would mean going from a life expectancy of about seventy-five to one hundred years or more of healthy, vigorous, active "old" age within our own generation. (Sorry about that, but our vocabulary is not keeping pace with our technology.) Already, as a recent article in the *New York Times* points out, we are seeing the advent of the "robust elderly," who are remarkably free of disease and disability as the rates of heart disease, stroke, arthritis, and emphysema have fallen dramatically (see Chapter 1). And let us assure you, this is just the beginning. The longer you can stick around, the greater the chances will be that your life span will roll out before you like sections of a highway that is in the process of being built. When that highway of life is completed, it will stretch to the ends of the earth as we know it.

SURFING THE AGE WAVE

Here are our predictions for the next fifty-five years for the advances in technology that will extend life span to virtual immortality in the coming millennium. Considering the pace at which anti-aging medical practice is growing, some of these predictions may occur even sooner than we anticipate.

In *five years*, based on current research projects nearing completion:

1: Cardiac treatments such as totally implantable artificial hearts (early versions of this device are already in use by some people), modified heart assist devices, and fetal cardiac cell transplantation that can grow new revitalized heart tissue will prolong the life spans of more than 40,000 Americans annually—just for starters. And this number could rise to several hundred thousand before the year 2010.

2: Implantable hormonal-pacemaker devices pumping out a concentrated mixture of growth hormone, other hormones, and growth factors in a cyclic rhythm that mimics the body's own pulsatile release will result in up to a thirty-year age reversal in the elderly.

3: Blood tests that screen for biochemical markers of DNA damage and early cancer growth will lead to detection of most cancers at a time when they are more than 90 percent curable.

4: Biomarker assays of the rate of human aging will allow anti-aging specialists to monitor each patient's progress in slowing the speed of biological aging and enable researchers to pick out the drug or therapy that is most effective in inhibiting or reversing the aging process. This advance should increase the average human life span in the United States to eighty-five years.

5: The ongoing experiments in diet restriction in monkeys, and now starting in human volunteers, will elucidate the underlying mechanisms of aging and point the way toward a doubling of the average life span that is now attained in experimental animals.

In *ten years*, based on the most advanced concepts of today:

1: The Human Genome Project, which has completed more than 90 percent of its goal of mapping all the human genes, will rocket genetic therapy to the forefront of clinical medicine. Genetic treatments will become standard for curing sickle cell anemia, metabolic disorders, and gene-based cancers.

2: Computerized telemedicine will bring televideo consultations into your own home, allowing access to top specialists from around the world for any medical disorder.

3: Average life span will increase to one hundred–plus years with no loss of mental function. Effective treatments for Alzheimer's disease and age-associated loss of memory and cognition will include "smart" drugs that improve mental processing speed, memory retention, and increase in IQ; embryonic cell implants to replace damaged neural tissue; and growth hormone, IGF–1, and growth factors to stimulate repair and regrowth of neurons within the patient's own brain.

In *thirty years*, the astounding progress will have revolutionized medical practice, making the stuff of today's science fiction tomorrow's reality. (Another reason to stick with the GH enhancement program.)

1: Nanotechnology, atom-by-atom construction, will yield molecule-sized instruments and computers much smaller than a human cell, allowing nanosurgeons to correct almost any biological malady or anatomical defect.

2: The human life span will average 150 years as genetic engineering eliminates virtually all genetic diseases and defects.

3: Backing up the brain will become a reality as people record their personality, memories, and consciousness on a microchip for the purpose of creating an immortal record of their psyches.

4: Bionic men and women will be a reality, as biomechanical prosthetic limbs that have the same dexterity but many times more strength than the original can be attached to the nerves and brain.

In *fifty-five years*, medical science will boldly go where no writer of *Star Trek* has ever gone before. (Speaking of the various *Star Trek*s, isn't it incredible how those twenty-fourth-century people go at warp speed through the galaxy and beam themselves through space but still grow older and look their age in earth years?)

1: Physical immortality will be a reality as the advent of human cloning permits total body replacement using either your original model or one designed to your specifications. The term "body shop" will take on new meaning.

2: Intellectual immortality will be possible through multiple memory psyche-chips about the size of a pinhead that will store the lifetime contents of an individual's thoughts, feelings, and memories.

BUILDING BRIDGES

While many of these predictions may strike you as fanciful, the science that will turn them into tangible realities is already being carried out in thousands of laboratories in this country and abroad—everything from biomarker testing to gene therapy to advanced drug delivery systems to cryopreservation of organs to cloning of embryonic cells to development of thinking machines to a new nano car, a working model of a Toyota car just produced by the Japanese that is the size of a grain of rice.

How will we span the gap from research to reality? Dr. Vincent Giampapa talks about the "bridging factor," anti-aging therapies that are available right now that will slow down the aging process enough so that those of us in the Baby Boomer age group and beyond can be healthy and fit enough to live another quarter century. At that point, science will have progressed enough to provide further bridges to extended health and life span.

Every great journey begins with a single step. Growth hormone, the first medically proven age-reversal therapy, is the beginning of humankind's greatest venture. It seems fitting that on the eve of a new millennium, we should be entering a new age of agelessness and limitless life span.

Welcome to the Ageless Society.

> Dr. Ronald Klatz
> Carol Kahn

Afterword

by Dr. Robert M. Goldman
President, National Academy of Sports Medicine

Human growth hormone is a powerful drug that has an immense potential for good in treating people whose pituitary glands release insufficient amounts. This includes the fastest growing segment of the population, those over sixty-five years of age. The use of growth hormone and other innovative substances promises to dramatically change the lives of millions.

But there is a shadow side to growth hormone, of which I am only too well aware—the possibility for abuse as a performance-enhancing drug. In 1984 Dr. Ronald Klatz and I published an exposé that blew the lid off the anabolic steroid epidemic among athletes, *Death in the Locker Room—Steroids and Sports*. The book became the bible for steroid education of athletes and was our statement that there is proper use of drugs for medical indications and there is abuse, such as solely for performance enhancement in normal hormonally balanced young people.

As chief medical officer and chairman of the International Medical Commission, which is the world governing federation for the sport of bodybuilding, I then went on to establish dope control and drug-testing programs for the sport for over 160 nations, working with Olympic Laboratories around the globe and with Dr. Manfred Donike, the man who developed the drug testing techniques for anabolic steroids that are still the state of the art around the world.

Unfortunately, some athletes, in their never-ending quest for the drug that will make them winners regardless of the cost to their health and bodies, began using dangerously large doses of growth hormone, which can't be detected in a urine test—the only test used in athletic competition. The dosages of HGH they used are many, many times that injected for hormone replacement purposes. At that level, they can produce all the symptoms of acromegaly, including premature death.

To prevent the use of human growth hormone in people for whom it is medically indicated because it might be abused by unscrupulous athletes is to throw out the baby with the bathwater. As Dr. Klatz and Carol Kahn note in this book, HGH is approved for use in the most vulnerable members of the population—young children. If it is safe for them, it seems to be very promising in older people who are unable to produce healthful amounts of growth hormone.

This book is the first major work for the general public detailing the huge amount of data and research in the field. Dr. Ronald Klatz, the central author, is the innovator behind the anti-aging movement, and I have had the pleasure of working closely with him for over fifteen years. He has this nasty talent of being so far ahead of the curve and the public mindset that unfortunately brings with it the pleasure of dealing with those critical of new innovations and dynamic change. His collaborator, Carol Kahn, lived, ate, and slept growth hormone, and did a remarkable job of pulling together the endless reams of data, research, and interviews. The authors are still strongly opposed to improper abuse of hormone medications for young hormonally balanced individuals, but very encouraged and intrigued by the potential good these powerful drugs may do for those deficient in natural production capability. They also show how people can stimulate their own endogenous levels of growth hormone through a program of dietary manipulation, exercise, and growth hormone releasers.

This book was a monumental task, and a courageous one. But in time hormone replacement for aging will be as commonplace as taking vitamins and lifting weights—which were also considered outlandish ideas in the not too distant past. One would have to bet that a lot of people will proactively pursue their own health quest and this book is an exceptional view of what the future may hold for all of us.

Resources for HGH and
Anti-Aging Therapies

Longevity Institute International
87 Valley Road
Montclair, NJ 07042
(201) 746-3533
Fax (201) 746-4385

The most comprehensive anti-aging clinic in America, offering a full range of diagnostic anti-aging measurements, HGH and hormone replacement therapeutics, anti-aging plastic surgery, and age-reversing medical interventions. This multi-specialty medical center is supervised by Dr. Vincent Giampapa.

Edwin M. Lichten, M.D.
29355 Northwestern Highway
Southfield MI 48034
(810) 350-3433
Fax (810) 358-2513
http://www.usdoctor.com

Dr. Lichten maintains a medical practice specializing in advanced forms of endocrinological therapies for resistant medical disorders including andropause and chronic migraine headache pain. He is an expert in the use of HGH and hormone replacement therapies for anti-aging purposes.

Meta Resorts
419 Park Avenue South, 4th Floor
New York, NY 10016
(800) 993-5559
Fax (212) 213-9049

A network of international health and healing spa facilities
managed by a group of board-certified physicians. These facilities
are designed from the ground up to provide a pampered resort spa
experience with a healthy dose of cutting edge technology and anti-
aging medical care, including HGH and other hormone replacement therapies, all under strict medical control.

Medical Center Pharmacy
Richard F. Fura, R.Ph.
10721 Main Street
Fairfax, VA 22030
(800) 723-7455
Fax (800) 238-8239

This compounding pharmacy offers a wide variety of both cus-
tom compounded hormonal replacement prescriptions and specific
anti-aging medications. The staff is extremely knowledgeable and
up-to-date on the latest advancements in drug therapies for aging.

International Anti-Aging Systems
P.O. Box 2995
London N10 2NA
England
Phone/Fax (UK) 181-444-8272

An international mail-order pharmacy specializing in longevity,
enhancement, and prevention. They provide high-quality anti-
aging hormone supplements and pharmaceuticals very economically, and their service and delivery time are very impressive.

Physicians Who Specialize in Anti-Aging Medicine

The following list includes physicians who practice advanced preventive medicine and anti-aging as a subspecialty.

Arthur Balin, M.D., Ph.D., F.A.C.P.
Center for Rejuvenation of
   Aging Skin
110 Chesley Campus
Media, PA 19063
(610) 892-0300

E.W. McDonagh, D.O.
McDonagh Medical Center, Inc.
2800A Kendallwood Parkway
Kansas City, MO 64119
(816) 453-5940

Allan Ahlschier, M.D.
8800 Starcrest Drive, Suite 168
San Antonio, TX 78217
(210) 653-2708

Dharma Singh Khalsa, M.D.
Alzheimer's Prevention Foundation
11901 East Coronado
Tucson, AZ 85749
(520) 749-8374

L. Stephen Coles, M.D., Ph.D.
Gerontology Research Group
4737 C LaVilla Marina
Marina del Rey, CA 90292-7037
(310) 827-3920
Fax (310) 827-0773
e-mail: scoles@grg.com

David Steenblock, M.S., D.O.
The Health Restoration Center
26381 Crown Valley Pkwy.
Suite 130
Mission Viejo, CA 92691
(714) 367-8870

Chong Park, M.D.
Plastic Surgery Center
89 Valley Road
Montclair, NJ 07042
(201) 746-3535

Vincent Giampapa, M.D.,
   F.A.C.S.
Plastic Surgery Center
89 Valley Road
Montclair, NJ 07042
(201) 746-3535

Ronald Hoffman, M.D.
The Hoffman Center
40 East 30th Street, 10th Floor
New York, NY 10016
(212) 779-1744

Gregory Keller, M.D., F.A.C.S.
2323 De La Vina Street
Santa Barbara, CA 93105
(805) 687-6408

Stephen Sinatra, M.D., F.A.C.C.,
   P.C.
Optimum Health
483 West Middle Turnpike,
   Suite 309
Manchester, CT 06040
(860) 643-5101

L. Terry Chappell, M.D.
Celebration of Health
122 Thurman Street, Box 248
Bluffton, OH 45817-0248
(419) 358-4627, (800) 788-4627

Binyamin Rothstein, D.O.
2835 Smith Avenue, Suite 208
Baltimore, MD 21209
(410) 484-2121

Barry DiBernardo, M.D.,
   F.A.C.S.
87 Valley Road
Montclair, NJ 07042
(201) 509-2000

Chris Renna, D.O.
The Renna Clinic
2705 Hospital Boulevard,
   Suite 206
Grand Prairie, TX 75051
(214) 641-6660
or
The Renna Clinic
1245 Sixteenth Street, Suite 307
Santa Monica, CA 90404
(800) 460-1959

Clinique La Prairie
Ch 1815
Clarens-Montreux
Switzerland
011-41-21-964-3311

Michael Perring, M.D.
Optimal Health of Harley Street
14 Harley Street
London W1N 1AG
011-171-935-5651

Eric Braverman, M.D.
Princeton Associates for Total
   Health
212 Commons Way, Building #2
Princeton, NJ 08540
(609) 921-1842

Jonathan Wright, M.D.
Tahoma Clinic
515 West Harrison, Suite 200
Kent, Washington 98032
(206) 854-4900
Fax (206) 850-5639

Julian Whitaker, M.D.
Whitaker Wellness Institute
4321 Birch Street, Suite 100
Newport Beach, CA 92660
(714) 851-1550

Thierry Hertoghe, M.D., Ph.D.
Avenue de l'Armee 127
1040 Bruxelles, Belgium
011-32-2-736-68-68
Fax: 011-32-2-732-5743

L. Cass Terry, M.D., Ph.D.
Froedtert Hospital
9200 West Wisconsin Avenue
Milwaukee, WI 53226
(414) 454-5204
Fax: (414) 259-0469

Elmer Cranton, M.D.
503 First Street, South, Suite 1
Yelm, WA 98597-7510
(360) 458-1061
Fax: (360) 894-3548

Karlis Ullis, M.D.
Sports Medicine/Anti-Aging
  Medicine
1454 Cloverfield Blvd., Suite 200
Santa Monica, CA 90404
(310) 829-1990
e-mail: kullis@ucla.edu

Steven Balbert, M.D.
1442 Ashbourne Road
Wyncote, PA 19095
(215) 886-7842

Roman Rozencwaig, M.D., C.M.
Westmount Medical Building
5025 Sherbrooke Street West,
  Suite 355
Montreal, Quebec H4A 1S9,
  Canada
(514) 487-0439
Fax (514) 487-0267
e-mail: romanroz@netcom.ca

R. Arnold Smith, Jr., M.D.
North Central Mississippi
  Regional Cancer Center
1401 River Road
Greenwood Leflore Hospital
Greenwood, MS 38930
(800) 720-8933

Pharmaceutical Manufacturers of HGH

Ferring (Germany)
400 Rella Boulevard, Suite 201
Suffern, NY 10901
(800) 445-3690

Pharmacia-Kabi (Sweden)
Box 16529
Columbus, OH 43216-6529
(614) 764-8100

Biotechnology-General (Israel)
70 Wood Avenue South
Iselen, NJ 08830
(908) 632-8800

Eli Lilly Research Labs (USA)
Indianapolis, IN 46285
(800) 545-5979

Genentech (USA)
460 Point San Bruno Boulevard
South San Francisco, CA 94080
(800) 821-8590

Novo Nordisk
100 Overlook Center, Suite 200
Princeton, NJ 08450
(609) 987-5800

Serano Labs, Inc.
100 Longwater Circle
Norwell, MA 02061
(617) 982-9000

Suppliers of GH-Releasers, Secretagogues, and Anti-Aging Nutrients

The suppliers listed offer the following:
A. Human Growth Hormone
B. Growth Hormone Releasing Hormone
C. Secretagogues
D. Anti-aging nutrients and nutritional GH-releasers

Amni Advanced Medical
    Nutrition (D)
Hayward, CA
(800) 437-8888

Biosyn (C, D)
5700 Citrus Blvd.
Hanrahan, LA 70125
(500) 734-7311, (800) 798-7809
Fax (500) 734-7333

Gero Vita International (B, C, D)
DC 9200
6021 Yonge Street
Toronto, Ontario M2M 3W2
Canada
(800) 406-1309

International Anti-Aging
    Systems (A, B)
P.O. Box 2995 M
N10 2NA England
011-181-444-8272

Laboratories de Longevite (A, B,
    C, D)
4840 Chemin Cote St. Luc Road,
    Suite 905
Montreal, Quebec HW3 2H1
Canada
(514) 486-8889
Fax (514) 486-5947

Life Extension Foundation (D)
P.O. Box 1097-K
Hollywood, FL 33022
(800) 333-2590
e-mail: growth@lef.org
Website: http://growth.lef.org.

Pan-Pacific Pharmacy (A, B, C, D)
Good Health Distributors
Tannery Block Ruby Industrial
    Complex
35 Tannery Road, No. 04-01
Singapore 1334

Medical Center Pharmacy (A, B)
10721 Main Street
Fairfax, VA 22030
(800) 723-7455

Weider Nutrition (C, D)
1960 South 4250 West
Salt Lake City, UT
(800) 436-3444

Optimum Health International,
    L.L.C. (D)
Manchester, CT
(800) 228-1507

Wellness Health & Pharmacy,
    Inc. (A, B)
Birmingham, AL
(800) 227-2627

Neither the authors nor the American Academy of Anti-Aging Medicine endorse or recommend any particular supplier or pharmacy listed herein. Caution should be used in all medical matters and the products provided by the above suppliers should be used only under the supervision of a licensed physician.

American Academy of
Anti-Aging Medicine Survey

Hormone replacement and other anti-aging medical therapies may be able to extend your potential healthy life span by thirty years or more! Help us to help you by completing this short survey. We will place you on our fax newsletter or e-mail list for updates on the latest breakthroughs in anti-aging medicine.

Age_____
Take vitamins daily_____
Exercise_____times a week.
Eat_____servings of fruits and vegetables/week.
_____months since last complete physical/medical exam

On a scale of 1 (poor) to 10 (excellent) rate the following:

General overall health_____
Satisfaction with life_____
Satisfaction with career_____
Satisfaction with relationships_____

Are you on an anti-aging program? Y/N_____

How did you like our book and what topics would you like to read about in the newsletter?

Name_____
Address_____
City, State, Zip_____
Fax_____ E-mail_____

Visit our Website: http://www.worldhealth.net
Send survey results to:

American Academy of Anti-Aging Medicine
1341 West Fullerton
Suite 111
Chicago, IL 60614

Bibliography

CHAPTER 1

Brody, J. "Restoring Ebbing Hormones May Slow Aging." *The New York Times*, July 18, 1995.

Hayflick, L. *How and Why We Age*. New York: Ballantine Books, 1996.

Hodes, R.J. "Frailty and Disability: Can Growth Hormone or Other Trophic Agents Make a Difference?" *Journal of the American Geriatrics Society* 42 (1994): 1208–11.

Jorgensen, J.O.L., et al. "Three Years of Growth Hormone Treatment in Growth Hormone-Deficient Adults: Near Normalization of Body Composition and Physical Performance." *European Journal of Endocrinology* 130 (1994): 224–28.

Kent, S. *The Life-Extension Revolution*. New York: Quill, 1980.

Klatz, R.M., ed. *Advances in Anti-Aging Medicine*, vol. 1. New York: Mary Ann Liebert, 1996.

Kolata, G. "New Era of Robust Elderly Belies the Fears of Scientists." *New York Times*, February 27, 1996.

Lewin, D.L. "Growth Hormone and Age: Something to Sleep On?" *Journal of NIH Research* 7 (1993): 34–35.

Rosen, T., G. Johannsson, et al. "Consequences of Growth Hormone Deficiency in Adults and the Benefits and Risks of Recombinant Human Growth Hormone Treatment." *Hormone Research* 43 (1995): 93–99.

Rudman, D., et al. "Effects of Human Growth Hormone in Men Over 60 Years Old." *New England Journal of Medicine* 323 (1990): 1–6.

Silverman, H.M., ed. *The Pill Book*, 6th edition. New York: Bantam Books, 1994.

Vreeland, L. "The Drug of the Decade?" *Ladies' Home Journal*, October 1990, 91–92.

Weiss R. "A Shot at Youth." *American Health*, November-December 1993.

CHAPTER 2

Bengtsson, B-A., et al. "Treatment of Adults with Growth Hormone (GH) Deficiency with Recombinant Human GH." *Journal of Clinical Endocrinology and Metabolism* 76 (1993): 309–17.

Christiansen, J.S., et al. "Effects of Growth Hormone on Body Composition in Adults." *Hormone Research* 33, suppl. 4 (1990): 61–64.

Erickson, D. "Big-Time Orphan: Human Growth Hormone Could Be a Blockbluster." *Scientific American*, September 1990, 165–66.

Falkeden, T. "Pathophysiological Studies Following Hypophysectomy in Man." Thesis, University of Gothenburg, 1963.

Fradkin, J., et al. "Creutzfeldt-Jakob Disease in Pituitary Growth Hormone Recipients in the United States." *Journal of the American Medical Association* (February 20, 1991): 880–84.

Harvey, S., C.G. Scanes, and W.H. Daughaday, eds. *Growth Hormone.* Boca Raton, Fla.: CRC Press, 1995.

Hodes, R.J. "Frailty and Disability: Can Growth Hormone or Other Trophic Agents Make a Difference?" *Journal of the American Geriatrics Society* 42 (1994): 1208–11.

Raben, M.S. "Treatment of Pituitary Dwarf with Human Growth Hormone." *Journal of Clinical Endocrinology* 18 (1958): 901–03.

Rosen, T., and B-A. Bengtsson. "Premature Mortality Due to Cardiovascular Disease in Hypopituitarism." *Lancet* 336 (1990): 285–88.

Rosen, T., I. Bosaeus, J. Tolli, G. Lindstedt, and B-A. Bengtsson. "Increased Body Fat Mass and Decreased Extracellular Fluid Volume in Adults with Growth Hormone Deficiency." *Clinical Endocrinology* 38 (1993): 63–71.

Rosen, T., et al. "Decreased Psychological Well-Being in Adult Patients with Growth Hormone Deficiency." *Clinical Endocrinology* 40 (1994): 111–16.

Rudman, D. "Growth Hormone, Body Composition, and Aging." *Journal of the American Geriatrics Society* 33 (1985): 800–07.

Rudman, D., et al. "Effects of Human Growth Hormone in Men Over 60 Years Old." *New England Journal of Medicine* 323 (1990): 1–6.

Sönksen, P.H. "Replacement Therapy in Hypothalamo-Pituitary Insufficiency After Childhood: Management in the Adult." *Hormone Research* 33, suppl. 4 (1990): 45–51.

Weiss, R. "A Shot at Youth." *American Health*, November-December 1993.

CHAPTER 3

Papadakis, M.A., et al. "Growth Hormone Replacement in Healthy Older Men Improves Body Composition But Not Functional Ability." *Annals of Internal Medicine* 124 (1996): 708–16.

Rudman, D., et al. "Effects of Human Growth Hormone in Men Over 60 Years Old." *New England Journal of Medicine* 323 (1990): 1–6.

Terry, L.C., and E. Chein, E. "Effects of Human Growth Hormone Administration (Low Dose-High Frequency) in 202 Patients." Personal communication.

Terry, L.C., and E. Chein, E. "Effects of Human Growth Hormone Administration (Low Dose-High Frequency) on Somatomedin C Blood Levels." Personal communication.

CHAPTER 4

Abribat, T., et al. "Alteration of Growth Hormone Secretion in Aging: Peripheral Effects." In *Growth Hormone II, Basic and Clinical Aspects*, edited by B.B. Bercu and R.F. Walker. New York: Springer-Verlag, 1994.

Ascoli, M., and D.L. Segaloff, D.L. "Adenohypophyseal Hormones and Their Hypothalamic Releasing Factors." In Goodman and Gilman's *The Pharmacological Basis of Therapeutics*, 9th edition. New York: McGraw-Hill, 1996.

Bengtsson, B-A. *An Introduction to Growth Hormone Deficiency in Adults.* Oxford, England: Oxford Clinical Communications, 1993.

Besser, G.M., and A.G. Cudworth eds., *Clinical Endocrinology: An Illustrated Text.* London: Gower Medical Publishing, 1987.

Curtis, H. *Biology*, 4th edition. New York: Worth Publishers, 1986.

Diamond, M.C., A.B. Schneibel, and L.M. Elson. *The Human Brain Coloring Book.* New York: Harper & Row, 1985.

Dilman, V.M. *The Grand Biological Clock.* Moscow: Mir, 1989.

Hertoghe, T. "Growth Hormone Therapy in Aging Adults: Place and Dosage in a Multiple Hormonal Replacement Therapy." 1996 European Congress on Anti-Aging Medicine, Madrid, Spain.

Iranmanesh, A., G. Lizarralde, and J.D. Veldhuis. "Age and Relative Adiposity Are Specific Negative Determinants of the Frequency and Amplitude of Growth Hormone (GH) Secretory Bursts and the Half-Life of Endogenous GH in Healthy Men." *Journal of Clinical Endocrinology and Metabolism* 73 (1991): 1081–88.

Jorgensen, J.O.L. "Adult Growth Hormone Deficiency." *Hormone Research* 42 (1994): 235–41.

Lehninger, A.L. *Principles of Biochemistry*. New York: Worth Publishers, 1982.

Rudman, D., et al. "Effects of Human Growth Hormone in Men Over 60 Years Old." *New England Journal of Medicine* 323 (1990): 1–6.

Shetty, K.R., and E.H. Duthie, Jr. "Anterior Pituitary Function and Growth Hormone Use in the Elderly." *Endocrinology and Metabolism Clinics of North America* 24, no. 2 (1995): 213–31.

Sonntag, W.E., et al. "Moderate Caloric Restriction Alters the Subcellular Distribution of Somatostatin mRNA and Increases Growth Hormone Pulse Amplitude in Aged Animals." *Neuroendocrinology* 61, no. 5 (1995): 601–08.

CHAPTER 5

Altman, L.K. "Powerful Response Reported to a Combined AIDS Therapy." *New York Times*, July 11, 1996.

Bjorksten, J. "The Crosslinkage Theory of Aging: Clinical Implications." *Comprehensive Therapy* 2 (1976): 65–74.

Chang, E., and C.B. Harley. "Telomere Length and Replicative Aging in Human Vascular Tissues." *Proceedings of the National Academy of Sciences USA* 90 (1995): 11190–94.

DaCosta, A.D. "Moderate Caloric Restriction Increases Type 1 IGF Receptors and Protein Synthesis in Aging Rats." *Mechanisms of Ageing Development* 71 (1993): 59–71.

Dilman, V.M. *The Grand Biological Clock*. Moscow: Mir, 1989.

Ettinger, Bruce, et al. "Reduced Mortality Associated with Long-Term Postmenopausal Estrogen Therapy." *Obstetrics and Gynecology* 87 (January 1996): 6–12.

Harman, D. "The Aging Process: Major Risk Factor for Disease and Death." *Proceedings of the National Academy of Sciences USA* 88 (June 1991): 5360–5363.

Hayflick, L. *How and Why We Age*. New York: Ballantine Books, 1996.

Hayflick, L. "The Limited In Vitro Lifetime of Human Diploid Cell Strains." *Experimental Cell Research* 37 (1965): 614–36.

Kahn, C. *Beyond the Helix: DNA and the Quest for the Secrets of Aging*. New York: Times Books, 1985.

Kelley, K.W., et al. "GH3 Pituitary Adenoma Implants Can Reverse Thymic Aging." *Proceedings of the National Academy of Sciences USA* 83 (1986): 5663–67.

Khansari, D.N., and T. Gustad. "Effects of Long-Term, Low-Dose Growth Hormone Therapy on Immune Function and Life Expectancy of Mice." *Mechanisms of Ageing and Development* 57 (1991): 87–100.

Weindruch, R. "Caloric Restriction and Aging." *Scientific American*, January 1966: 46–52.

Weindruch, R., and R.L. Walford. *The Retardation of Aging and Dietary Restriction*. New York: Charles C. Thomas, 1988.

Wong, G. Personal communication.

CHAPTER 7

Bar-Dayan, Y., and M. Small. "Effect of Bovine Growth Hormone Administration on the Pattern of Thymic Involution in Mice." *Thymus* 23, no. 2 (1994): 95–101.

Crist, D.M., et al. "Exogenous Growth Hormone Treatment Alters Body Composition and Increases Natural Killer Cell Activity in Women with Impaired Endogenous Growth Hormone Secretion." *Metabolism* 36, no. 12 (1987): 1115–17.

Davila, D.R., et al. "Role of Growth Hormone in Regulating T-Dependent Immune Events in Aged, Nude, and Transgenic Rodents." *Journal of Neuroscience Research* 18 (1987): 108–16.

Goff, B.L., et al. "Growth Hormone Treatment Stimulates Thymulin Production in Aged Dogs." *Clinical and Experimental Immunology* 68 (1987): 580–97.

Goya, R.G., M.C. Gagnerault, M.C. De Moraes, W. Savino, and M. Dardeen. "In Vivo Effects of Growth Hormone on Thymus Function in Aging Mice." *Brain Behav. Immun.* 6, no. 4 (December 1992): 341–54.

Harvey, S. "Growth Hormone Action: Neural Function." In *Growth Hormone*, edited by S. Harvey, C.G. Scanes, and W.H. Daughaday. Boca Raton, Fla.: CRC Press, Boca Raton, 1995.

Kelley, K.W. "Growth Hormone in Immunobiology." In *Psychoneuroimmunology*, 2nd edition, edited by R. Ader, D.L. Feltern, and N. Cohen. New York: Academic Press, 1991.

Kelley, K.W. "Immunologic Roles of Two Metabolic Hormones, Growth Hormone and Insulin-Like Growth Factor-I, in Aged Animals." *Nutritional Reviews* 53, no. 4 (1995): S95–S104.

Kelley, K.W., et al. "GH3 Pituitary Adenoma Implants Can Reverse Thymic Aging." *Proceedings of the National Academy of Sciences USA* 83 (1986): 5663–67.

Kelley K.W., et al. "A Pituitary-Thymus Connection During Aging." *Annals of the New York Academy of Sciences* 521 (1988): 88–98.

Khansari, D.N., and T. Gustad. "Effects of Long-Term, Low-Dose Growth Hormone Therapy on Immune Function and Life Expectancy of Mice." *Mechanisms of Ageing and Development* 57 (1991): 87–100.

Lawler, B., et al. "Growth Hormone Treatment Stimulates Thymulin Production in Aged Dogs." *Clinical and Experimental Immunology* 68 (1987): 580–87.

Sonntag, W.E., et al. "Moderate Caloric Restriction Alters the Subcellular Distribution of Somatostatin mRNA and Increases Growth Hormone Pulse Amplitude in Aged Animals." *Neuroendocrinology* 61, no. 5 (1995): 601–08.

Weigent, D.A., and J.E. Blalock. "Immunoreactive Growth Hormone-Releasing Hormone in Rat Leukocytes." *Journal of Neuroimmunology* 29 (1990): 1–13.

CHAPTER 8

Angelin, B. and M. Rudling. "Growth Hormone and Lipoprotein Metabolism." *Endocrinology and Metabolism* 2, suppl. B (1995): 25–28.

Bak, B., P.H. Jorgensen, and T.T. Andreassen. "The Stimulating Effect of Growth Hormone on Fracture Healing Is Dependent on Onset and Duration of Administration." *Clinical Orthopedics* 264, March 1991: 295–301.

Beaubien, G. "Why Get Old?" *Chicago Tribune*, October 20, 1994.

Bengtsson, B-A. "The Consequences of Growth Hormone Deficiency in Adults." *Acta Endocrinologica* 128, suppl. 2 (1993): 2–5.

Bengtsson, B-A. "Effects of Bone Mineral Content; Stimulation of Bone and Cartilage Growth." Submitted for publication.

Bengtsson, B-A. *An Introduction to Growth Hormone Deficiency in Adults.* Oxford, England: Oxford Clinical Communications, 1993.

Bengtsson, B-A., et al. "Cardiovascular Risk Factors in Adults with Growth Hormone Deficiency." *Endocrinology and Metabolism* 2, suppl. B (1995): 29–35.

Bengtsson, B-A., et al. "Treatment of Adults with Growth Hormone (GH) Deficiency with Recombinant Human GH." *Journal of Clinical Endocrinology and Metabolism* 76 (1993): 309–17.

Beshyah, S.A. "The Effects of Short and Long Term Growth Hormone Replacement Therapy in Hypopituitary Adults on Lipid Metabolism and Carbohydrate Tolerance." *Journal of Clinical Endocrinology and Metabolism* 80 (1995): 356–63.

Bier, D.M. "Growth Hormone and Insulin-Like Growth Factor I: Nutritional Pathophysiology and Therapeutic Potential." *Acta-Paediatr.-Scand.-Suppl.* 374 (1991): 119–28.

Brixen, K., et al. "Growth Hormone (GH) and Adult Bone Remodeling: The Potential Use of GH in Treatment of Osteoporosis." *Journal of Pediatric Endocrinology* (England) 6, no. 1 (January-March 1993): 65–71.

Brownlee, S., and T. Watson. Can a Hormone Fight AIDS? (Human Growth Hormone)." *U.S. News & World Report*, November 21 1994.

Caidahl, K., S. Eden, and B-A. Bengtsson. "Cardiovascular and Renal Effects of Growth Hormone, *Clinical Endocrinology* (Oxford, England) 40, no. 3 (March 1994): 393–400.

Christensen, H., H. Oxlund, and S. Laurberg. "Postoperative Biosynthetic Human Growth Hormone Increases the Strength and Collagen Deposition of Experimental Colonic Anastomoses." *International Journal of Colorectal Diseases* 6, no. 3 (August 1991): 133–38.

Christiansen, J.S., et al. "GH-Replacement Therapy in Adults." *Hormone Research* 36, suppl. 1 (1991): 66–72.

Cole, B.R. "Recombinant Human Growth Hormone Therapy in Renal Insufficiency: New Hope for Children" (editorial). *Journal of the American Society of Nephrology* 1, no. 10 (April 1991): 1127.

Corpas, E., S.M. Harman, and M.R. Blackman. "Human Growth Hormone and Human Aging." *Endocrine Reviews* 14, no. 1 (1993): 20–39.

Cuneo, R.C., et al. "Cardiac Failure Responding to Growth Hormone." *Lancet*, April 15, 1989: 838–9.

Cuneo, R.C., et al. "Cardiovascular Effects of Growth Hormone Treatment in Growth-Hormone-Deficient Adults: Stimulation of the Renin-Aldosterone System." *Clinical Sciences* 81, no. 5 (November 1991): 587–92.

Daubeney, P.E.F., et al. "Cardiac Effects of Growth Hormone in Short Normal Children: Results After Four Years of Treatment." *Archives of Disease in Childhood* 72 (1993): 337–39.

De Boer, H.D., G.J. Blok, and E.A. Van Der Veen. "Clinical Aspects of Growth Hormone Deficiency in Adults." *Endocrine Reviews* 16, no. 1 (1995): 63–86.

Degerblad, M., et al. "Reduced Bone Mineral Density in Adults with Growth Hormone Deficiency: Increased Bone Turnover During 12 Months of GH Substitution Therapy." *European Journal of Endocrinology* 133, no. 2 (August 1995): 180–88.

Eriksen, E.F., M. Kassem, and K. Brixen. "Growth Hormone and Insulin-Like Growth Factors As Anabolic Therapies for Osteoporosis." *Hormone Research* 40, nos. 1-3 (1993): 95–98.

Falanga, V. "Growth Factors and Wound Healing." *Dermatologic Clinics* 11, no. 4 (October 1993): 66–75.

Fazio, S. "Preliminary Study of Growth Hormone in the Treatment of Dilated Cardiomyopathy." *New England Journal of Medicine* 334 (March 28, 1996): 809–14.

Frustaci, A., et al. "Reversible Dilated Cardiomyopathy Due to Growth Hormone Deficiency." *American Journal of Clinical Pathology* 97 (1992): 503–11.

Gottardis, M., A. Benzer, W. Koller, T.J. Luger, F. Puhringer, and J. Hackl. "Improvement in Septic Syndrome After Administration of Recombinant Human Growth Hormone (rhGH)." *Journal of Trauma* 31, no. 1 (January 1991): 81–86.

Hesse, V., et al. "Insulin-Like Growth Factor I Correlations to Changes of the Hormonal Status in Puberty and Age." *Experimental and Clinical Endocrinology* 102, no. 4 (1994): 289–98.

Imagawa, S., M.A. Goldberg, J. Doweiko, and H.F. Bunn. "Regulatory Elements of the Erythropoietin Gene." *Blood* 77, no. 2 (January 1991): 278–85.

Inzucchi, S.E., and R.J. Robbins. "Effects of Growth Hormone on Human Bone Biology." *Journal of Clinical Endocrinology and Metabolism* 79, no. 3 (1994): 691–94.

James, J.S. "Human Growth Hormone Reverses Wasting in Clinical Trial." *AIDS Treatment News*, August 19, 1994.

Johannsson, G., et al. "Two Years of Growth Hormone Treatment in Growth-Hormone-Deficient Adults Normalised Body Composition and Induced Favourable Changes in Cardiovascular Risk Factors." In press.

Johannsson, G., T. Rosen, I. Bosaeus, L. Sjorstrom, and B-A. Bengtsson. "Two Years of Growth Hormone Treatment Increases Bone Mineral

Content and Density in Hypopituitary Patients with Adult-Onset Growth Hormone Deficiency." *Journal of Clinical Endocrinology and Metabolism* 81, no. 8 (1996): 2865–73.

Johannsson, J.O., L. Kerstin, L. Tengborn, T. Rosen, and B-A. Bengtsson. "High Fibrinogen and Plasminogen Activator Inhibitor Activity in Growth Hormone-Deficient Adults." Manuscript in preparation.

Kaiser, F.E., A.J. Silver, and J.E. Morley. "The Effect of Recombinant Human Growth Hormone on Malnourished Older Individuals." *Journal of the American Geriatrics Society* 39 (1991): 235–40.

Kaplan, S. L. "The Newer Uses of Growth Hormone in Adults." *Advances in Internal Medicine* 38 (1993): 287ff.

Kimbrough, T.D., S. Sherman, T.R. Ziegler, M. Scheltinga, and D.W. Wilmore. "Insulin-Like Growth Factor-I Response Is Comparable Following Intravenous and Subcutaneous Administration of Growth Hormone." *Journal of Surgical Research* 51 (December 1991): 472–76.

Kovacs, G., et al. "Growth Hormone Prevents Steroid-Induced Growth Depression in Health and Uremia." *Kidney-Int.* 40, no. 6 (December 1991): 1032–40.

Loh, E., and J.L. Swain. "Growth Hormone for Heart Failure—Cause for Cautious Optimism." *New England Journal of Medicine* 334 (March 28, 1996): 856–57.

Mulligan, K., C. Grunfeld, M.K. Hellerstein, R.A. Neese, and M. Schambelan. "Anabolic Effects of Recombinant Human Growth Hormone in Patients with Wasting Associated with Human Immunodeficiency Virus Infection." *Journal of Clinical Endocrinology and Metabolism* 77 (1993): 956–62.

Ohlsson, C., et al. "Growth Hormone Induces Multiplication of the Slowly Cycling Germinal Cells of the Rat Tibial Growth Plate." *Proceedings of the National Academy of Science USA* 89 (October 1992): 9826–30.

Powrie, J., A. Weissberger, and P. Sonksen. "Growth Hormone Replacement Therapy for Growth Hormone-Deficient Adults." *Drugs* 49, no. 5 (1995): 656–62.

Rosen, T., et al. "Cardiovascular Risk Factors in Adult Patients with Growth Hormone Deficiency." *Acta Endocrinologica* 129, no. 3 (September 1993): 195–200.

Rosen, T., et al. "Consequences of Growth Hormone Deficiency in Adults and the Benefits and Risks of Recombinant Human Growth Hormone Treatment." *Hormone Research* 43 (1995): 93-99.

Rosen, T., G. Johannsson, and B-A. Bengtsson. "Consequences of Growth Hormone Deficiency in Adults, and Effects of Growth Hormone Replacement Therapy." *Acta Paediatrica* suppl. 399 (1994): 21–24.

Rubin, C.D. "Southwestern Internal Medicine Conference: Growth Hormone—Aging and Osteoporosis." *American Journal of Medical Science* 305, no. 2 (1993): 120–29.

Rudling, M., et al. "Importance of Growth Hormone for the Induction of Hepatic Low Density Lipoprotein Receptors." *Proceedings of the National Academy of Sciences USA* 89 (August 1992): 6983–87.

Salva, P.S., and G.E. Bacon. "Public Interest in Human Growth Hormone Therapy" (letter). *Western Journal of Medicine* 155, no. 4 (October 1991): 428.

Takagi, K., T. Tashiro, Y. Mashima, H. Yamamori, K. Okui, and I. Ito. "The Effect of Human Growth Hormone on Protein Metabolism in the Surgically Stressed State" (in Japanese). *Nippon-Geka-Gakkai-Zasshi* 92, no. 11 (November 1991): 1545–51.

Valcavi, R., O. Gaddi, M. Zini, M. Lavicol, U. Mellino, and I. Portioli. "Cardiac Performance and Mass in Adults with Hypopituitarism: Effects of One Year of Growth Hormone Treatment." *Journal of Clinical Endocrinology and Metabolism* 80, no. 2 (1995): 659ff.

Vergani, G., A. Mayerhofer, and A. Bartke. "Acute Effects of Human Growth Hormone on Liver Cells in Vitro: A Comparison with Livers of Mice Transgenic for Human Growth Hormone." *Tissue-Cell.* 23, no. 5 (1991): 607–12.

Voerman, B.J. "Effects of Human Growth Hormone in Critically Placebo-Controlled Trial." *Critical Care Medicine* 23, no. 4 (1995): 665–73.

Walker, R.F., G.C. Ness, Z. Zhao, and B.B. Bercu. "Effects of Stimulated Growth Hormone Secretion on Age-Related Changes in Plasma Cholesterol and Hepatic Low Density Lipoprotein Messenger RNA Concentrations." *Mechanisms of Ageing and Development* 73, no. 3 (September 1994): 215–26.

Yang, R., S. Bunting, N. Gillett, R. Clark, and H. Jin. "Growth Hormone Improves Cardiac Performance in Experimental Heart Failure." *Circulation* 92 (1995): 262–67.

Ziegler, T.R., J.M. Lazarus, L.S. Young, R. Hakim, and D.W. Wilmore. "Effects of Recombinant Human Growth Hormone in Adults Receiving Maintenance Hemodialysis." *Journal of the American Society of Nephrology* 2, no. 6 (1991): 1130–5.

CHAPTER 9

Bengtsson, B-A. *An Introduction to Growth Hormone Deficiency in Adults*. Oxford, England: Oxford Clinical Communications, 1993.

Bengtsson, B-A., et al. "Treatment of Adults with Growth Hormone (GH) Deficiency with Recombinant Human GH." *Journal of Clinical Endocrinology and Metabolism* 76 (1993): 309–17.

Beshyah, S.A. "The Effects of Short and Long Term Growth Hormone Replacement Therapy in Hypopituitary Adults on Lipid Metabolism and Carbohydrate Tolerance." *Journal of Clinical Endocrinology and Metabolism* 80 (1995): 356–63.

Christiansen, J.S., et al. "Effects of Growth Hormone on Body Composition in Adults." *Hormone Research* 33, suppl. 4 (1990): 61–64.

Clemmons, D.R., and L.E. Underwood. "Growth Hormone As a Potential Adjunctive Therapy for Weight Loss." In *Human Growth Hormone: Progresses and Challenges*, edited by L.E. Underwood. New York: Marcel Dekker, 1986.

Cordido, F., et al. "Massive Growth Hormone Discharge in Obese Subjects After the Combined Administration of GH-Releasing Hormone and GHRP–6: Evidence for a Marked Somatotroph Secretory Capability in Obesity." *Journal of Clinical Endocrinology and Metabolism* 76 (1983): 819–23.

Jorgensen, J.O.L. "Adult Growth Hormone Deficiency." *Hormone Research* 42 (1994): 235–41.

Jorgensen, J.O.L., et al. "Three Years of Growth Hormone Treatment in Growth Hormone Deficient Adults: Near Normalization of Body Composition and Physical Performance." *European Journal of Endocrinology* 130 (1994): 224–8.

Orme, S.M. "Comparison of Measures of Body Composition in a Trial of Low Dose Growth Hormone Replacement Therapy." *Clinical Endocrinology* 37 (1992): 453–59.

Ottosson, M., et al. "Growth Hormone Inhibits Lipoprotein Lipase Activity in Human Adipose Tissue." *Journal of Clinical Endocrinology and Metabolism* 80 (1995): 936–41.

Papadakis, M.A., et al. "Growth Hormone Replacement in Healthy Older Men Improves Body Composition But Not Functional Ability." *Annals of Internal Medicine* 124 (1996): 708–16.

Richelsen, B., et al. "Growth Hormone Treatment of Obese Women for 5 Weeks: Effect on Body Composition and Adipose Tissue LPL Activity." *American Journal of Physiology* 226, no. 2, pt. 1 (1994): E211–16.

Rosenbaum, M., J.M. Gertner, and R. Leibel, R. "Effects of Systemic Growth Hormone Administration on Regional Adipose Tissue Distribution and Metabolism in GH-Deficient Children." *Journal of Clinical Endocrinology and Metabolism* 69 (1989): 1274–81.

Rudman, D. "Growth Hormone, Body Composition, and Aging." *Journal of the American Geriatrics Society* 33 (1985): 800–07.

Rudman, D. "Impaired Growth Hormone Secretion in the Adult Population: Relation to Age and Adiposity." *Journal of Clinical Investigation* 67 (1981): 1361–69.

Rudman, D., A.G. Feller, L. Cohn, K.R. Shetty, and I.W. Rudman. "Effects of Human Growth Hormone on Body Composition in Elderly Men." *Hormone Research* 36, suppl. 1 (1991): 73–81.

Rudman, D., et al. "Effects of Human Growth Hormone in Men Over 60 Years Old." *New England Journal of Medicine* 323 (1990): 1–6.

Rudman, D., et al. "Relations of Endogenous Anabolic Hormones and Physical Activity to Bone Mineral Density and Lean Body Mass in Elderly Men." *Clinical Endocrinology* 40 (1994): 653–61.

Russell-Jones, D.L. "The Effects of Growth Hormone on Protein Metabolism in Adult Growth Hormone Deficient Patients." *Clinical Endocrinology* 38 (1993): 427–31.

Salomon, F., R.C. Cuneo, R. Hesp, and P.H. Sonksen. "The Effects of Treatment with Recombinant Human Growth Hormone on Body Composition and Metabolism in Adults with Growth Hormone Deficiency." *New England Journal of Medicine* 321 (1989): 1797–803.

Snyder, D., L.D. Underwood, and D.R. Clammons. "Persistent Lipolytic Effect of Exogenous Growth Hormone During Caloric Restriction." *American Journal of Medicine* 98 (1995): 129–34.

Weindruch, R. "Caloric Restriction and Aging." *Scientific American*, January 1966: 46–52.

Weindruch, R., and R.L. Walford. *The Retardation of Aging and Dietary Restriction.* New York: Charles C. Thomas, 1988.

Whitehead, et al. "Growth Hormone Treatment of Adults with Growth Hormone Deficiency: Results of a 13-Month Placebo Controlled Cross-Over Study." *Clinical Endocrinology* 36 (1992): 45–52.

CHAPTER 10

Bengtsson, B-A., et al. "Treatment of Adults with Growth Hormone (GH) Deficiency with Recombinant Human GH." *Journal of Clinical Endocrinology and Metabolism* 76 (1993).

Corpas, Emiliano, S.M. Harman, and M.R. Blackman."Human Growth Hormone and Human Aging." *Endocrine Reviews* 14, no. 1 (1993): 20–39.

Daughaday, W.H., and S. Harvey. "Growth Hormone Action: Clinical Significance." In *Growth Hormone*, edited by S. Harvey, C.G. Scanes, and W.H. Daughaday. Boca Raton, Fla.: CRC Press, Boca Raton, 1995.

Falanga, Vincent. "Growth Factors and Wound Healing." *Dermatologic Clinics* 11, no. 4 (October 1993): 667-675.

Ho, K.Y., and A.J. Weussberger. "The Antinatriuretic Action of Biosynthetic Human Growth Hormone in Man Involves Activation of the Renin-Angiotensin System." *Metabolism* 39 (1990): 133–37.

Rosen, T., I. Bosaeus, J. Tolli, G. Lindstedt, and B-A. Bengtsson. "Increased Body Fat Mass and Decreased Extracellular Fluid Volume in Adults with Growth Hormone Deficiency." *Clinical Endocrinology* 38 (1993): 63–71.

Savard, P. "Growth Factors and Rejuvenation: General Overview of the Therapeutical Uses of GHs and Their Market." Paper presented at the Third International Conference on Anti-Aging Medicine and Biomedical Technology, Las Vegas, Nevada, December 9–11, 1995.

CHAPTER 11

Abraham, I. "Potency Insurance: Powerful Plant Medicines and Nutrients Can Aid Lifetime Virility and Potency." *Journal of Longevity Research* 1, no. 11 (1996).

Libby, R. "The Glow of Desire." *Journal of Longevity Research* 1, no. 11 (1996).

Perring, M., and J. Moral. "Holistic Approach to the Management of Erectile Disorders in a Male Sexual Health Clinic." *British Journal of Clinical Practice* 49, no. 3 (May-June 1995): 140.

Rosen, T., et al. "Decreased Psychological Well-Being in Adult Patients with Growth Hormone Deficiency." *Clinical Endocrinology* 40 (1994): 111–16.

CHAPTER 12

Bjorntorp, P. "Neuroendocrine Ageing." *Journal of Internal Medicine* 238, no. 5 (November 1995): 401–04.

Deijen, J.B., et al. "Cognitive Impairments and Mood Disturbances in Growth Hormone Deficient Men." Manuscript in preparation.

Hoffman, A.R., et al. "Growth Hormone Therapy in the Elderly: Implications for the Aging Brain." *Psychoneuroimmunology* 17, no. 4 (1992): 327–333.

Johannsson, J.O., G. Larson, M. Andersson, A. Elmgren, L. Hynsjo, A. Lindahl, P.A. Lundberg, O.G.P. Isaksson, S. Lindstedt, and B-A. Bengtsson. "Treatment of Growth Hormone-Deficient Adults with Recombinant Human Growth Hormone Increases the Concentration of Growth Hormone in the Cerebrospinal Fluids and Affects Neurotransmitters." *Neuroendocrinology* 61 (1995): 57–66.

Johnston, B.M., E.C. Mallard, C.E. Williams, and P.D. Gluckman. "Insulin-Like Growth Factor–1 Is a Potent Neuronal Rescue Agent After Hypoxic-Ischemic Injury in Fetal Lambs." *Journal of Clinical Investigation* 97, no. 2 (January 15, 1996): 300–08.

McGauley, G.A., et al. "Psychological Well-Being Before and After Growth Hormone Treatment in Adults with Growth Hormone Deficiency." *Hormone Research* 33, suppl. 4 (1990): 52–4.

Libby, R. "The Glow of Desire." *Journal of Longevity Research* 1, no. 11 (1996).

Perring, M., and J. Moral. "Holistic Approach to the Management of Erectile Disorders in a Male Sexual Health Clinic." *British Journal of Clinical Practice* 49, no. 3 (May-June 1995): 140.

Rosen, T., et al. "Decreased Psychological Well-Being in Adult Patients with Growth Hormone Deficiency." *Clinical Endocrinology* 40 (1994): 111–16.

Sartorio, A., E. Molinari, G. Riva, A. Conti, F. Morabitom and G. Falia. "Growth Hormone Treatment in Adults with Childhood Onset Growth Hormone Deficiency: Effects on Psychological Capabilities." *Hormone Research* 44, no. 1 (1995): 6–11.

Shetty, K.R., and E.H. Duthie, Jr. "Anterior Pituitary Function and Growth Hormone Use in the Elderly." *Endocrinology and Metabolism Clinics of North America* 24, no. 2 (1995): 213–31.

Sonntag, W.E., et al. "L-Dopa Restores Amplitude of Growth Hormone Pulses in Old Male Rats to That Observed in Young Male Rats." *Neuroendocrinology* 34 (1982): 163–68.

Warwick D.J., D.B. Lowrie, and P.J. Cole. "Growth Hormone Activation of Human Monocytes for Superoxide Production But Not Tumor Necrosis Factor Production, Cell Adherence, or Action Against Mycobacterium Tuberculosis." *Infection and Immunology* 63, no. 11 (November 1995): 4312–16.

CHAPTER 13

Bengtsson, B-A. *An Introduction to Growth Hormone Deficiency in Adults.* Oxford, England: Oxford Clinical Communications, 1993.

Cuneo, R.C., et al. "Diagnosis of Growth Hormone Deficiency." *Clinical Endocrinology* 37 (1992): 387–97.

Dean, W. *Biological Aging Measurement: Clinical Applications.* Los Angeles: The Center for Bio-Gerontology, 1988.

De Boer, H., et al. "Diagnosis of Growth Hormone Deficiency in Adults." *Lancet* 343 (June 25, 1994): 1645–46.

Hoffman, D.M., et al. "Diagnosis of Growth-Hormone Deficiency in Adults." *Lancet* 343 (1994): 1064–68.

Landin-Wilheimsen, et al. "Serum Insulin-Like Growth Factor I in a Random Population Sample of Men and Women: Relation to Age, Sex, Smoking Habits, Coffee Consumption and Physical Activity, Blood Pressure and Concentrations of Plasma Lipids, Fibrinogen, Parathyroid Hormone and Osteocalcin." *Clinical Endocrinology* 41 (1994): 351–57.

Walford, R.L. *The 120-Year Diet.* New York: Simon and Schuster, 1986.

CHAPTER 14

Giampapa, V.C., R.M. Klatz, B.E. Di Bernardo, and F.A. Kovarik. "Biomarker Matrix Protocol." In *Advances in Anti-Aging Medicine*, vol. 1, edited by R.M. Klatz. New York: Mary Ann Liebert, 1996.

CHAPTER 15

Andrews, W.C. "Continuous Combined Estrogen/Progestin Hormone Replacement Therapy." *Hospital Medicine*, supplement (November 1995): 1–11.

Apgar, B., et al. "Menopause: Should Hormone-Replacement Therapy Be Routine?" *Female Patient* 20 (May 1995): 39–54.

Baker, B. "Estrogen May Be Effective for Stroke Reduction." *Family Practice News*, November 1, 1995, 16.

Baker, B. "HRT May Guard Against Some Cancer Recurrence." *Family Practice News*, November 1, 1995, 16.

Barrett-Connor, E., and D. Goodman-Gruen. "The Epidemiology of DHEAS and Cardiovascular Disease." In *Dehydroepiandrosterone (DHEA) and Aging*. Edited by F. Bellino, et al. *Annals of the New York Academy of Sciences* 774 (1995): 259–70.

Bilger, B. "Forever Young." *Sciences*, September-October 1995, 27–31.

Birenhager-Gillesse, E.G., J. Derksen, and A.M. Lagaay. "Dehydroepiandrosterone Sulphate (DHEAS) in the Oldest Old, Age 85 and Older." *Annals of the New York Academy of Sciences* 719 (1994): 543–52.

Brenner, D.E., et al. "Postmenopausal Estrogen Replacement Therapy and the Risk of Alzheimer's Disease: A Population-Based Case Control Study." *American Journal of Endocrinology* 140, no. 3 (1994): 262–67.

Brody, J.E. "Experimental Evidence Is Lacking for Melatonin As Cure-All." *New York Times*, September 27, 1995, C9.

Colgan, M. *Hormonal Health*. Vancouver, British Columbia: Apple Publishing, 1996.

Combes, A. "Balancing Act." *New York Times Magazine*, March 3, 1996, 31.

Cowley, G. "Melatonin." *Newsweek*, August 7, 1995, 46–49.

DeVita, E. "The Hormone Craze." *American Health* (January-February 1996): 73.

"DHEA Replacement Therapy." *Life Extension Report* 13, no. 9 (September 1993): 65–72.

Ditkoff, E.C., W.G. Crary, M. Cristo, et al. "Estrogen Improves Psychological Function in Asymptomatic Postmenopausal Women." *Obstetrics and Gynecology* 78, no. 6 (December 1991): 991–95.

Douglass, W.C. "A Neglected Hormone—Testosterone for Men and Women, Part 1." *Second Opinion* 5, no. 3 (March 1995): 28–29.

Ettinger, B., et al. "Reduced Mortality Associated with Long-Term Postmenopausal Estrogen Therapy." *Obstetrics and Gynecology* 87 (1996): 6–12.

Friend, T. "Study May Calm Breast Cancer Estrogen Fears." *USA Today*, March 26, 1996.

Gaby, A.R. "DHEA: The Hormone That Does It All." *Holistic Medicine* (Spring 1993): 19–23.

Gibbs, R.B. "Estrogen and Nerve Growth Factor-Related Systems in Brain. Effects on Basal Forebrain and Cholinergic Neurons and Implications for Learning and Memory Processes and Aging." *Annals of the New York Academy of Sciences* 743 (November 14, 1994): 165–96.

Gordon, L. "Removal of Ovaries Poses Cardiovascular Risks, Study Shows." *Medical Tribune*, December 7, 1995, 4.

Gorman, C. "Lost Fountain of Youth." *Time*, February 5, 1996, 53.

Hamalainen, E., H. Adlercreutz, P. Puska, et al. "Diet and Serum Sex Hormones in Healthy Men." *Journal of Steroid Biochemistry* 20, no. 1 (1984): 459–64.

Hargrove, J., W. Maxon, A. Wentz, et al. "Menopausal Hormone-Replacement Therapy with Continuous, Daily, Oral, Micronized Estradiol and Progesterone." *American Journal of Obstetrics and Gynecology* 73 (1989): 606.

Henderson, V. W., A. Paganini-Hill, C. K. Emanuel, et al. "Estrogen Replacement Therapy in Older Women: Comparisons Between Alzheimer's Disease Cases and Nondemented Control Subjects." *Archives of Neurology* 51 (September 1994): 896–900.

Ho, K.Y., et al. "Effects of Sex and Age on the 24-Hour Profile of Growth Hormone Secretion in Man: Importance of Endogenous Estradiol Concentrations." *Journal of Clinical Endocrinology and Metabolism* 64, no. 1 (1987): 51–8.

Hobbs, C.J., et al. "Testosterone Administration Increases Insulin-like Growth Factor-1 Levels in Normal Men." *Journal of Clinical Endocrinology and Metabolism* 77 (1993): 776–779.

Janowsky, J.S., S.K. Oviatt, E.S. Orwoll. "Testosterone Influences Spatial Cognition in Older Men." *Behavior Neuroscience* 108, no. 2 (April 1994): 325–32.

Kampen, D.L., and B.B. Sherwin. "Estrogen Use and Verbal Memory in Healthy Postmenopausal Women." *Obstetrics Gynecology* 83, no. 6 (June 1994): 979–83.

Klatz, R.M. "The Clock Hormone." *Journal of Longevity Research* 1, no. 9 (1995): 38–40.

Klatz, R.M. and R. Goldman. *Stopping the Clock*. New Canaan, CT: Keats Publishing, 1996.

Kritz-Silverman, D., and E. Barrett-Connor. "Long-Term Postmenopausal Hormone Use, Obesity, and Fat Distribution in Older Women." *Journal of the American Medical Association* 275 (January 3, 1996): 46–49.

Langer, S.E., and J.F. Scheer. *Solved: The Riddle of Illness*. New Canaan, Conn.: Keats Publishing, 1984.

Lissoni, P., A. Ardizzoia, S. Barni, et al. "Efficacy and Tolerability of Cancer Neuroimmunotherapy with Subcutaneous Low-Dose Interleukin-2 and the Pineal Hormone Melatonin: A Progress Report of 200 Patients with Advanced Solid Neoplasms." *Oncology Reports* 2 (1995): 1064–68.

Liu, L., G.R. Merriam, R.J. Sherins, et al. "Chronic Steroid Exposure Increases Mean Plasma Growth Hormone Concentration and Pulse Amplitude in Men with Isolated Hypogonadotropic Hypogonadism." *Journal of Clinical Endocrinology and Metabolism* 64, no. 4 (1987): 651–56.

Matsumoto, A.M. "Andropause—Are Reduced Androgen Levels in Aging Men Physiologically Important?" *Western Journal of Medicine* 159, no. 5 (November 1993): 618–20.

Mokshagundam, S.P., and U. Barzel. "Thyroid Disease in the Elderly." *Journal of the American Geriatrics Society* 41 (1993): 1361–69.

Monmaney, T., and S. Katz. "A User's Guide to Hormones." *Newsweek*, January 12, 1987, 50–56.

Morales, A.J., J.J. Nolan, S.S.C. Yen, et al. "Effects of Replacement Dose of Dehydroepiandrosterone in Men and Women of Advancing Age." *Journal of Clinical Endocrinology and Metabolism* 78, no. 6 (1994): 1360–67.

Paganini-Hill, A. "The Benefits of Estrogen Replacement Therapy on Oral Health/The Leisure World Cohort." *Archives of Internal Medicine* 25, no. 155 (November 1995): 2325–29.

Painter, K. "Postmenopausal Drugs Pass FDA Hurdles." *USA Today*, December 4, 1995, 4D.

Pardridge, W.M., R.A. Gorski, B.M. Lippe, et al. "Androgens and Sexual Behavior." *Annals of Internal Medicine* 96 (1982): 488–501.

Peat, R. "Thyroid: Misconceptions." *Townsend Letter for Doctors* November 1993, 1120–22.

Perring, M. "Reviews: Biochemistry." *Care of the Elderly* (February 1994): 58–59.

Peters, S.L. "Some Women on Fast Track Feel Derailed." *USA Today*, February 1, 1996, 1A.

Ratloff, J. "Drug of Darkness: Can a Pineal Hormone Head Off Everything from Breast Cancer to Aging?" *Science News* 147 (May 13, 1995): 300–1.

Regelson, W., and M. Kalimi. "Dehydroepiandrosterone (DHEA)—A Pleiotropic Steroid: How Can One Steroid Do So Much?" In R.M. Klatz, ed., *Advances in Anti-Aging Medicine*, vol. 1, Larchmont, NY: Mary Ann Liebert, Inc., 1996.

Reiter, R.J. and J. Robinson. *Melatonin*. New York: Bantam Books, 1995.

Robinson, D., L. Friedman, R. Marcus, et al. "Estrogen Replacement Therapy and Memory in Older Women." *Journal of the American Geriatrics Society* 42, no. 9 (1994): 919–22.

Rozencwaig, R., B.R. Grad, and J. Ochoa. "The Role of Melatonin and Serotonin in Aging." *Medical Hypotheses* 23, no. 4 (August 1987): 337.

Sack, M., D.J. Rader, and R.O. Cannon. "Oestrogen and Inhibition of Oxidation of Low-Density Lipoproteins in Postmenopausal Women." *Lancet* 343 (January 29, 1994): 269–70.

Sahelian, R. "Melatonin, the Natural Sleep Medicine." *Total Health* 17 (August 1995): 30–2.

Sherwin, B.B. "Estrogenic Effects on Memory in Women." *Annals of the New York Academy of Sciences* 743 (November 14, 1994): 213–30.

Stanford, J.L., et al. "Combined Estrogen and Progestin Hormone Replacement Therapy in Relation to Risk of Breast Cancer in Middle-Aged Women." *Journal of the American Medical Association* 274 (July 12, 1995): 137–142.

Tamkins, T. "Hormone-Replacement Therapy May Prevent Tooth Loss." *Medical Tribune*, January 11, 1996, 19.

Tamkins, T. "Hormone-Replacement Therapy Raises Breast-Cancer Risk." *Medical Tribune*, July 13, 1995, 4.

Tenover, J.S. "Androgen Administration to Aging Men." *Clinical Andrology* 23, no. 4 (1994): 877–87.

T.F.P. "Hormone Replacement Therapy in Menopause." *Female Patient* 18 (April 1993): 99–100.

Vermeulen, A. "Dehydroepiandrosterone Sulfate and Aging." In Dehydroepiandrosterone (DHEA) and Aging. Edited by F. Bellino, et al. *Annals of the New York Academy of Sciences* 774 (1995): 121–7.

Whitaker, J. *Dr. Julian Whitaker's Health and Healing* 2, no. 16 (June 1992): 2–3.

Whitaker, J. *Dr. Julian Whitaker's Health and Healing* 4, no. 1 (February 1994): 4–5.

Wilshire, G.B., J.S. Loughlin, J.R. Brown, et al. "Diminished Function of the Somatotropic Axis in Older Reproductive-Aged Women." *Journal of Clinical Endocrinology and Metabolism* 80, no. 1 (1995): 608–13.

Wilson, J.O. and J.E. Griffiri. "The Use and Misuse of Androgens." *Metabolism* 29, no. 12 (December 1980): 1278–89.

Writing Group for PEPI Trial. "Effects of Estrogen or Estrogen/Progestin Regimens on Heart Disease Risk Factors in Postmenopausal Women: The Postmenopausal Estrogen/Progestin Interventions (PEPI) Trial." *Journal of the American Medical Association* 273, no. 3 (January 18, 1995): 199–208.

Yen, S.S.C., A.J. Morales, and O. Khorram. "Replacement of DHEA in Aging Men and Women." In *Dehydroepiandrosterone (DHEA) and Aging*. Edited by F. Bellino, et al. *Annals of the New York Academy of Sciences* 774 (1995): 128–42.

CHAPTER 16

Alba-Roth, J., O.A. Muller, J. Schopohl, et al. "Arginine Stimulates Growth Hormone Secretion By Suppressing Endogenous Somatostatin Secretion." *Journal of Clinical Endocrinology and Metabolism* 67, no. 6 (1988): 1186–89.

Borst, J.E., et al. "Studies of GH Secretagogues in Man." *Journal of the American Geriatrics Society* 42, no. 5 (May 1995): 532–4.

Bunt, J.C., T.A. Boileau, J.M. Bahr, et al. "Sex and Training Differences in Human Growth Hormone Levels During Prolonged Exercise." *Journal of Applied Physiology* 61, no. 5 (November 1986): 1796–801.

Calabresi, P., et al. "l-Deprenyl Test in Migraine: Neuroendocrinological Aspects." *Cephalagia* 13 (1993): 406–09.

Caroni, P. "Activity Sensitive Signaling by the Muscle Derived Insulin-Like Growth Factor in the Developing and Regenerating Neuromuscular System." *Annals of the New York Academy of Sciences* 692 (1993): 201–08.

Casanueva, F.F., L. Villaneuva, J.A. Cabranes, et al. "Cholinergic Mediation of Growth Hormone Secretion Elicited by Arginine, Clonidine, and Physical Exercise in Man." *Journal of Clinical Endocrinology and Metabolism* 52, no. 3 (March 1981): 409–15.

Cash, C.D. "Gammahydroxybutyrate: An Overview of the Pros and Cons for It Being a Neurotransmitter And/Or a Useful Therapeutic Agent." *Neuroscience and Biobehavioral Reviews* 18, no. 2 (1994): 291–304.

Ceda, G.P., G. Ceresini, G. Denti, et al. "Alpha Glycerylphosphorylcholine Administration Increases the GH Responses to GHRH of

Young and Elderly Subjects." *Hormone Metabolism Research* 24 (1992): 119–21.

Cella, S.G., Valerio Moiraghi, Francesco Minuto, et al. "Prolonged Fasting or Clonidine Can Restore the Defective Growth Hormone Secretion in Old Dogs." *Acta Endocrinologica* (Copenhagen) 121, no. 2 (1989): 177–84.

Chalmers, R.J., and R.H. Johnson. "The Effect of Diphenylhydatoin on Metabolic and Growth Hormone Changes During and After Exercise." *Journal of Neurology, Neurosurgery, and Psychiatry* 46, no. 7 (July 19837): 662–65.

Chaney, M.M. "The Effect of Oral Arginine, Age, and Exercise on Growth Hormone, Insulin, and Blood Glucose." *Dissertation Abstracts International* 50, no 10 (April 1990).

Coiro, V., et al. "Reduction of Baclofen-, But Not Sodium Valproate-Induced Growth Hormone Release in Type 1 Diabetic Men." *Hormone Metabolism Research* 23 (1991): 600–04.

Conteras, V. "Natural Method for Boosting Human Growth Hormone." *Journal of Longevity Research* 1, no. 8 (1995): 38–39.

Corpas, E., M.R. Blackman, R. Roberson, et al. "Oral Arginine-Lysine Does Not Increase Growth Hormone or Insulin-Like Growth Factor–1 in Old Men." *Journal of Gerontology* 48, no. 4 (1993): M128–133.

D'Alessandro, R., et al. "Phenytoin-induced Increase in Growth Hormone Response to Levodopa in Adult Males." *Journal of Neurology, Neurosurgery, and Psychiatry* 47 (1984): 715–19.

Durso, R., C.A. Tamminga, A. Denaro, et al. "Plasma Growth Hormone and Prolactin Response to Dopaminergic Gabamimetic and Cholinergic Stimulation in Huntington's Disease." *Neurology* 33 (September 1983): 1229–32.

Froesch, E.R.; Zenobi, P.D.; Hussain, M. "Metabolic and Therapeutic Effects of Insulin-Like Growth Factor 1, *Hormone Research* (1994)42: 66–71.

Gann, H., et al. "Growth-Hormone Response to Clonidine in Panic Disorder Patients in Comparison to Patients with Major Depression and Healthy Controls." *Pharmacopsychiatry* 28, no. 3 (May 1995): 80–83.

Ghigo, E., et al. "Arginine Abolishes the Inhibitory Effect of Glucose on the Growth Hormone Response to Growth Hormone-Releasing Hormone in Man." *Metabolism* 141, no. 9 (September 1992): 1000–03.

Ghigo, E., et al. "Arginine Potentiates the GHRH-But Not the Pyri-dostigmine-Induced GH Secretion in Normal Short Children. Further Evidence for a Somatostatin Suppressing Effect of Arginine." *Clinical Endocrinology* 32 (1990): 763–67.

Ghigo, E., et al. "Growth Hormone (GH) Responsiveness to Combined Administration of Arginine and GH-Releasing Hormone Does Not Vary with Age in Man." *Journal of Clinical Metabolism* 17, no. 6 (1990): 1481–85.

Hobbs, C.J., et al. "Effect of T Administration on IGF–1 Levels." *Journal of Clinical Endocrinology and Metabolism* 77, no. 3 (1993): 776–79.

Hoffman, W.H., J.T. DiPiro, R.L. Tackett, et al. "Relationship of Plasma Clonidine to Growth Hormone Concentrations in Children and Adolescents." *Journal of Clinical Pharmacology* 29, no. 6 (June 1989): 538–42.

Isidori, A., A.L. Monaco, M. Cappa, et al. "A Study of Growth Hormone Release in Man After Oral Administration of Amino Acids." *Current Medical Research and Opinion* 7, no. 7 (1981): 475–81.

Karpik, K., B. Miller, P. Frickner, et al. "Testosterone and Growth Hormone Responses to Hypnosis and Exercise—A Pilot Study." *New Zealand Journal of Sports Medicine* 15, no. 4 (December 1987: 88–91.

Koppeschaar, H.P.F., C.D. ten Horn, J.H.H. Thijssen, et al. Differential Effects of Arginine on Growth Hormone Releasing Hormone and Insulin Induced Growth Hormone Secretion." *Clinical Endocrinology* 36 (1992): 487–90.

Kupfer, S.R., L.E. Underwood, R.C. Baxter, et al. "Enhancement of the Anabolic Effects of Growth Hormone and Insulin-Like Growth Factor–1 by Use of Both Agents Simultaneously." *Journal of Clinical Investigation* 91, no. 2 (February 1993): 391–96.

Laurian L., et al. "Growth Hormone Response to L-Dopa in the Thinned Obese." *Israeli Journal of Medical Science* 18, no. 5 (May 1982: 625–29.

Luoma, P.V., et al. "Elevated Serum Growth Hormone Levels in Patients Treated with Anticonvulsants." In *Advances* in *Epileptology; XIth Epilepsy International Symposium*. Edited by R. Canger, F. Angeleri, and J.K. Peary. New York: Raven Press, 1980.

Martal, J., N. Chene, and P. de la Llosa. "Involvement of Lysin Residues in the Binding of Hgh and Bgh to Somatotropic Receptors." *Federation of European Biochemical Sciences* 180, no. 2 (1985): 295–99.

Matteini, M., et al. "GH Secretion by Arginine Stimulus: The Effect of Both Low Doses and Oral Arginine Administered Before Standard Test." *Bullettino Della Societa Italiana di Biologica Sperimentale* 56 (1980): 2254–61.

Mauras, N., and B. Beaufrere. "Recombinant Human Insulin-Like Growth Factor–1 Enhances Whole Body Protein Anabolism and Significantly Diminishes the Protein Catabolic Effects of Prednisone in Humans Without a Diabetogenic Effect." *Journal of Endocrinology and Metabolism* 80, no. 3 (1995): 869–74.

McCann, S.M., and V. Rettori. "Gamma Amino Butyric Acid (GABA) Controls Anterior Pituitary Hormone Secretion." In *GABA and Endocrine Function*. Edited by G. Racagni and A. Donoso. New York: Raven Press, 1986.

McCune, M.A., et al. "Treatment of Recurrent Herpes Simplex Infections with L-Lysine Monohydrochloride." *Cutis* 34 (1983): 366–73.

Munson, M. "Turn Back Time: Can This Nutrient Hold Off Aging?" *Prevention*, June 1995, 25–27.

Pavlov, E.P., M.S. Harman, G.R. Merriam, et al. "Responses of Growth Hormone (GH) and Somatomedin-C to GH Releasing Hormone in Healthy Aging Men." *Journal of Clinical Endocrinology and Metabolism* 62, no. 3 (1986): 595–600.

Pearson, D. and S. Shaw. "GH Releasers: An Update Review." *Life Extension Newsletter* 3, No. 2 (July-August 1990).

Prewett, N.A., P. Bettica, S. Mohan, et al. "Age-Related Decreases in Insulin-Like Growth Factor–1 and Transforming Growth Factor-B in Femoral Cortical Bone from Both Men and Women: Implications for Bone Loss with Aging." *Journal of Clinical Endocrinology and Metabolism* 78, no. 5 (1994): 1011–16.

Quabbe, Hans-Jurgen, Stephan Bunge, Thomas Walz, et al. "Plasma Glucose and Free Fatty Acids Modulate the Secretion of Growth Hormone, But Not Prolactin, in the Rhesus and Java Monkey." *Journal of Clinical Endocrinology and Metabolism* 70, no. 4 (1990): 908–15.

Reeds, P.J., K.A. Munday, and M.R. Turner. "The Effect of Growth Hormone in Vitro on Muscle Accumulation and Incorporation of Arginine." *Hormone Metabolism Research* 3 (1971): 129–30.

Rolandi, E., et al. "Changes of Pituitary Secretion After Long-Term Treatment with Hydergine, in Elderly Patients." *Acta Endocrinologica* 102 (1983): 332–36.

Rosen, C.J., et al. "Insulin-Like Growth Factors and Bone: The Osteoporosis Connection." *Proceedings of the Society for Experimental Biology* 206 (1994): 83–102.

Sato, T., et al. "Mutual Priming Effects of GHRH and Arginine on GH Secretion: Informative Procedure for Evaluating GH Secretory Dynamics." *Endocrinology, Japan* 37, no. 4 (1990): 501–09.

Steardo, L., M. Iovino, P. Monteleone, M. Argusta, et al. "Evidence for a GABAenergic Control of the Exercise-Induced Rise in GH in Man." *European Journal of Clinical Pharmacology* 28 (1985): 607–08.

Takahara, J., S. Yonuki, W. Yakushiji, et al. Stimulatory Effects of Gamma-Hydroxybutyric Acid on Growth Hormone and Prolactin Release in Humans." *Journal of Clinical Endocrinology and Metabolism* 44 (1977): 1014–17.

Uusitupa, M., O. Siitonen, M. Haerkoenen, et al. "Modification of the Metabolic and Hormonal Response to Physical Exercise by Beta-Blocking Agents." *Annals of Clinical Research* 14, suppl. 34 (19824): 165–67.

Welbourne, T. "Increased Plasma Bicarbonate and Growth Hormone After Oral Glutamine Load." *American Journal of Clinical Nutrition* 61 (1995): 1058–61.

Yarasheski, K.E., J.J. Zachweija, T.J. Angelopoulos, et al. "Short-Term Growth Hormone Treatment Does Not Increase Muscle Protein Synthesis in Experienced Weight Lifters." *Journal of Applied Physiology* 74, no. 6 (1995): 3073-76.

Zorgniotti, A.W., and E.F. Lizza. "Effect of Large Doses of the Nitric Oxide Precursor, L-Arginine, on Erectile Dysfunction." *International Journal of Impotence Research* 6 (1994): 33–36.

Zsuzsanna, A.C.S., G. Szabo, G. Kapocs, et al. "y-Aminobutyric Acid Stimulates Pituitary Growth Hormone Secretion in the Neonatal Rat, a Superfusion Study." *Endocrinology* 120, no. 5 (1987): 1790–98.

CHAPTER 17

Iranmanesh, A., B. Lizarralde, and J.D. Veldhuis. "Age and Relative Adiposity Are Specific Negative Determinants of the Frequency and Amplitude of Growth Hormone (GH) Secretory Bursts and the Half-Life of Endogenous GH in Healthy Men." *Journal of Clinical Endocrinology and Metabolism* 73 (1991): 1081–88.

Laritcheva, K.A., et al. "Study of Energy Expenditure and Protein Needs of Top Weight Lifters." In *Nutrition, Physical Fitness and Health*, edited by J. Parizkova and V.A. Rogozkin. Baltimore: University Park Press, 1978.

Lemon, P., et al. "The Importance of Protein for Athletes." *Sports Medicine* 1 (1984): 474–84.

Pearson, D., and S. Shaw. *The Life Extension Weight Loss Program*. New York: Doubleday, 1986.

Rabinowitz, D., and K.L. Ziebler. "A Metabolic Regulating Device Based on the Actions of Human Growth Hormone and of Insulin, Singly and Together, on the Human Forearm." *Nature* 199 (August 31, 1963): 913–15.

Sears, B., and B. Lawren. *The Zone*. New York: HarperCollins, 1995.

CHAPTER 18

Borst, S.E., et al. "Growth Hormone, Exercise and Aging: The Future of Therapy for the Frail Elderly." *Journal of the American Geriatrics Society* 42 (1994): 529–35.

Bucci, L.R., et al. "Effect of Ferulate on Strength and Body Composition of Weightlifters." *Journal of Applied Sport Science Research* 4, no. 3 (1990): 104.

Cuneo, R.C., et al. "Growth Hormone Treatment in Growth Hormone-Deficient Adults I: Effects on Muscle Mass and Strength." *Journal of Applied Physiology* 70 (1991): 688–94.

Cuneo, R.C., et al. "Growth Hormone Treatment in Growth Hormone-Deficient Adults II: Effects on Exercise Performance." *Journal of Applied Physiology* 70 (1991): 695–700.

Farrell, P.A., et al. "Influence of Endogenous Opioids on the Response of Selected Hormones to Exercise in Humans." *Journal of Applied Physiology* 61 (1986): 1051–57.

Goldfard, A.H., et al. "Plasma Beta-Endorphin Concentration: Response to Intensity and Duration of Exercise." *Medical Science of Sports Exercise* 22 (1990): 241–44.

Hagberg, J.M., et al. "Metabolic Responses to Exercise in Young and Older Athletes and Sedentary Men." *Journal of Applied Physiology* 65 (1988): 900–08.

Kozlovsky, A., et al. "Effects of Diets High in Simple Sugars on Urinary Chromium Losses." *Metabolism* 35, no. 6 (1986): 515–18.

Kraemer, W.J., et al. "Endogenous Anabolic Hormonal and Growth Factor Responses to Heavy Resistance Exercise in Males and Females." *International Journal of Sports Medicine* 12 (1991): 228–35.

Luger, A., et al. "Plasma Growth Hormone and Prolactin Responses to Graded Levels of Acute Exercise and to a Lactate Infusion." *Neuroendocrinology* 56 (1992): 112–17.

Nass, R., et al. "Effect of Growth Hormone (hGH) Replacement Therapy on Physical Work Capacity and Cardiac and Pulmonary Function in Patients with hGH Deficiency Acquired in Adulthood." *Journal of Clinical Endocrinology and Metabolism* 80 (1995): 552–7.

Paoletti, R., ed. *Drugs Affecting Lipid Metabolism.* New York: Elsevier Publishing Company, 1980.

Popov, I.M., and W.J. Goldway. "A Review of the Properties and Clinical Effects of Ginseng." *American Journal of Chinese Medicine* 1, no. 2 (1973): 263–70.

Pyka, G., et al. "Age-Dependent Effect of Resistance Exercise on Growth Hormone Secretion in People." *Journal of Clinical Endocrinology and Metabolism* (1992)75: 404–07.

Siliprandi, N., and M.T. Ramacci. "Carnitine As a "Drug" Affecting Lipid Metabolism." In *Coenzyme Q–10 and Clinical Aspects of Coenzyme 9*, edited by R. Fumagalli, D. Kritchevsky, J. Vanfranchem, and K. Folkers. Amsterdam: Biomedical Press, 1981.

VanHelder, W., et al. "Growth Hormone Regulation in Two Types of Aerobic Exercise of Equal Oxygen Uptake." *European Journal of Applied Physiology* 55 (1986): 236–39.

Ward, P.E. and R.D. Ward. *Encyclopedia of Weight Training.* Laguna Hills, CA: QPT Publications, 1991.

Weltman, A., et al. "Endurance Training Amplifies the Pulsatile Release of Growth Hormone: Effects of Training Intensity." *Journal of Applied Physiology* 72 (1992): 2188–96.

Wolff, B. "Barbell Exercises." *Muscle & Fitness*, August 1966, 112–15.

CHAPTER 19

Albertsson, W.K., et al. "Recombinant Somatropin in Treatment of Growth Hormone Deficient Children in Sweden and Finland." *Acta Paediatr. Scand.* 347, Suppl. (1988): 176–79.

Beshyah, S.A., V. Anyaoku, R. Niththyananthan, P. Sharp, and D.G. Johnston. "The Effect of Subcutaneous Injection Site on Absorption of Human Growth Hormone: Abdomen Versus Thigh." *Clinical Endocrinology Oxford* 25, no. 5 (November 1991): 409–12.

Bierich, J.R. "Multicentre Clinical Trial of Authentic Recombinant Somatropin Growth Hormone Deficiency." *Acta Paediatr Scand* 337, suppl. (1987): 135–40.

De Boer, H., G.J. Blok, B. Voerman, P. DeVries, C. Popp-Snijders, and E. Van Der Veen. "The Optimal Growth Hormone Replacement Dose in Adults, Derived from Bioimpedance Analysis." *Journal of Clinical Endocrinology and Metabolism* 80, no. 7 (1995): 2069–76.

Ho, K.Y., et al. "Effects of Sex and Age on the 24-Hour Profile of Growth Hormone Secretion in Man: Importance of Endogenous Estradiol Concentrations." *Journal of Clinical Endocrinology and Metabolism* 64, no. 1 (1987): 51.

Johannsson, G., et al. "The Individual Responsiveness to Growth Hormone (GH) Treatment in GH-Deficient Adults Is Dependent on the Level of GH-Binding Protein, Body Mass Index, Age, and Gender." *Journal of Clinical Endocrinology and Metabolism* 81 (1996) 81: 1–7.

Jorgensen, J.T., et al. "Development, Production and Pharmacodynamics of Human Growth Hormone," *Indian Journal of Pediatrics* (India) 58, suppl. 1 (September-October 1991): 23–32.

Kyriazis, M. "Role of Growth Hormone in Ageing." *British Journal of Clinical Practice* 49 (February 1994): 56–7.

Laursen, T., J.O. Jorgensen, G. Jakobsen, B.L. Hansen, and J.S. Christiansen. "Continuous Infusion Versus Daily Injections of Growth Hormone (GH) for 4 Weeks in GH-Deficient Patients." *Journal of Clinical Endocrinology and Metabolism* 80, no. 8 (August 1995): 2410–18.

Lehrman, Sally. "The Fountain of Youth? (Human Growth Hormone)." *Harvard Health Letter* 17, no. 3 (June 1992): 1.

Neely, E.K., and R.G. Rosenfeld. "Use and Abuse of Human Growth Hormone." *Annual Review of Medicine* 45 (1994): 407–20.

Rudman, D., A.G. Feller, L. Cohn, K.R. Shetty, and I.W. Rudman. "Effects of Human Growth Hormone on Body Composition in Elderly Men." *Hormone Research* 36, suppl. 1 (1991): 73–81.

Sobel, Dava. "Youth Revisited." *Genetic Engineering News* 10 (February 1991).

Stone, Richard. "NIH to Size Up Growth Hormone Trials." *Science* 257, no. 1 (August 7, 1992): 739,

Weiss, R. "A Shot at Youth." *American Health*, November-December 1993.

Zadik, Z., et al. "The Influence of Age on the 24-Hour Integrated Concentration of Growth Hormone in Normal Individuals." *Journal of Clinical Endocrinology and Metabolism* 60, no. 3 (1985): 513–16.

CHAPTER 20

AMA Council on Scientific Affairs. "Drug Abuse in Athletes: Anabolic Steroids and Human Growth Hormone." *Journal of the American Medical Association* 259 (March 18, 1988): 1703–05.

Fradkin, J., et al. "Risk of Leukemia After Treatment with Pituitary Growth Hormone." *Journal of the American Medical Association* 270 (December 15, 1993): 2829–32.

Goglia, P., and J. Kindela. "Out of Control: One Man's 15-Year Roller-Coaster Ride Propelled by Anabolic Steroids and Other Performance-Enhancing Drugs." *Flex* (January 1996).

Goldman, R., and R. Klatz. *Death in the Locker Room—Steroids and Sports*. Indianapolis: Harper/Icarus Press, 1984.

Holmes, S. and S.M. Shalet. "Factors Influencing the Desire for Long-Term Growth Hormone Replacement in Adults." *Clinical Endocrinology* 43 (1995): 151–57.

Holmes, S. and S.M. Shalet. "Which Adults Develop Side-Effects of Growth Hormone Replacement?" *Clinical Endocrinology* 43 (1995): 143–49.

Lehrman, S. "The Fountain of Youth? (Human Growth Hormone)." *Harvard Health Letter* 17, no. 3 (June 1992): 1.

Marcus, R. "Should All Older People Be Treated with Growth Hormone?" *Drugs & Aging* (January 8, 1996): 1–4.

Papadakis, M.A., et al. "Growth Hormone Replacement in Healthy Older Men Improves Body Composition But Not Functional Ability." *Annals of Internal Medicine* 124 (1996): 708–16.

Todd, T. "Growth Hormone." *Muscle & Fitness* (February 1995).

Weiss, R. "Human Growth Hormone Treatment-Leukemia Link Reported." *Science News* 133 (1993): 308.

Yarasheski, K.E., and J.J. Zachwieja. "Growth Hormone Therapy for the Elderly: The Fountain of Youth Proves Toxic" (letter to the editor). *Journal of the American Medical Association* 227 (October 13, 1993): 894.

CHAPTER 21

Bellone, J., et al. "Growth Hormone-Releasing Activity of Hexarelin, a New Synthetic Hexapeptide, Before and During Puberty." *Journal of Clinical Endocrinology and Metabolism* 80, no. 4 (1995): 1090–94.

Bowers, C.Y. "GH Releasing Peptides—Structure and Kinetics." *Journal of Pediatric Endocrinology* 6, no. 1 (1993): 21–31.

Bowers, C.Y. "An Overview of the GH Releasing Peptides (GHRPs), Growth Hormone Secretagogues" (abstract). Paper presented at Serano Symposia USA, December 8–11, 1994.

Bowers, C.Y., D.K. Alster, and J.M. Frentz. "The Growth Hormone-Releasing Activity of a Synthetic Hexapeptide in Normal Men and Short Statured Children After Oral Administration." *Journal of Clinical Endocrinology and Metabolism* 74 (1992): 292–98.

Bowers, C.Y., et al. "On the In Vitro and In Vivo Activity of a New Synthetic Hexapeptide That Acts on the Pituitary to Specifically Release Growth Hormone." *Endocrinology* 114, no. 5 (1984): 1537.

Bowers, C.Y., G.A. Reynolds, D. Durham, et al. "Growth Hormone (GH)-Releasing Peptide Stimulates GH Release in Normal Men and Acts Synergistically with GH-Releasing Hormone." *Journal of Clinical Endocrinology and Metabolism* 70, no. 4 (19904): 975–82.

Bowers, C.Y., A.O. Sartor, G.A. Reynolds, et al. "On the Actions of Growth Hormone-Releasing Hexapeptide, GHRP." *Endocrinology* 128, no. 4 (1991): 2027–35.

Chapman, I.M., et al. "Stimulation of the Growth Hormone (GH)/IGF-1 Axis by Daily Oral Administration of a GH Secretagogue (MK-0677) in Healthy Elderly Subjects." *Journal of Clinical Endocrinology and Metabolism* 81 (1966): 4249–57.

Chingo, et al. "Growth Hormone-Releasing Activity of Growth Hormone-Releasing Peptide–6 Is Maintained After Short-Term Oral Pretreatment with the Hexapeptide in Normal Aging." *European Journal of Endocrinology* 131, no. 5 (1994): 4899–503.

Cordido, F., et al. "Massive Growth Hormone Discharge in Obese Subjects After the Combined Administration of GH-Releasing Hormone and GHRP–6: Evidence for a Marked Somatotroph Secretory Capability in Obesity." *Journal of Clinical Endocrinology and Metabolism* 76 (1983): 819–23.

Corpas, E., et al. "Continuous Subcutaneous Infusions of Growth Hormone (GH) Releasing Hormone 1–44 for 14 Days Increase GH and Insulin-Like Growth Factor–1 Levels in Old Men." *Journal of Clinical Endocrinology and Metabolism* 76, no. 1 (1993): 134–38.

Corpas, E., et al. "Growth Hormone (GH)-Releasing Hormone-(1–29) Twice Daily Reverses the Decreased GH and Insulin-Like Growth Factor–1 Levels in Old Men." *Journal of Clinical Endocrinology and Metabolism* 75, no. 2 (1992): 530–35.

"Editorial: On a Peptidomimetic Growth Hormone Releasing Peptide." *Journal of Clinical Endocrinology and Metabolism* 79, no. 4 (1994): 940–42.

Froesch, E.R., P.D. Zenobi, and M. Hussain. "Metabolic and Therapeutic Effects of Insulin-Like Growth Factor 1, *Hormone Research* (1994) 42: 66–71.

Ghigo, E., E. Arvat, L. Gianotti, et al. "Growth Hormone-Releasing Activity of Hexarelin, a New Synthetic Hexapeptide, After Intravenous Subcutaneous, Intranasal, and Oral Administration in Man." *Journal of Clinical Endocrinology and Metabolism* 78, no. 3 (1994): 693–98.

Goth, Miklos I., Charles E. Lyons, Benedict J. Canny, et al. "Pituitary Adenylate Cyclase Activating Polypeptide, Growth Hormone (GH)-Releasing Peptide and GH-Releasing Hormone Stimulate GH Release Through Distinct Pituitary Receptors." *Endocrinology* 130, no. 2 (1992): 939–44.

Hansson, H.A. "Insulin-like Growth Factors and Nerve Regeneration." *Annals of the New York Academy of Sciences* 692 (1993): 161–171.

Hartman, M.L., et al. "Oral Administration of Growth Hormone (GH)-Releasing Peptides Stimulates GH Secretion in Normal Men." *Journal of Clinical Endocrinology and Metabolism* 74 (1992): 1378–84.

Howard, A.D., et al. "A Receptor in Pituitary and Hypothalamus That Functions in Growth Hormone Release." *Science* 273 (August 16, 1966): 974–77.

Jaffe, C.A., et al. "Effects of a Prolonged Growth Hormone (GH)-Releasing Peptide Infusion on Pulsatile GH Secretion in Normal Men." *Journal of Clinical Endocrinology and Metabolism* 77, no. 6 (1993): 1641–47.

Lieberman, S.A., et al. "Anabolic Effects of Recombinant Insulin-Like Growth Factor–1 in Cachectic Patients with the Acquired Immunodeficiency Syndrome." *Journal of Clinical Endocrinology and Metabolism* 78, no. 2 (1994): 404–10.

Lo, H.C., et al. "Simultaneous Treatment with IGH–1 and GH Additively Increases Anabolism in Parenterally Fed Rats." *American Journal of Physiology* 269, no. 2, pt. 1 (August 1995): 368–76.

Loche, S., et al. "The New Growth Hormone-Releasing Activity of Hexarelin, a New Synthetic Hexapeptide, in Short Normal and Obese Children and in Hypopituitary Subjects." *Journal of Clinical Endocrinology and Metabolism* 80, no. 2 (1995): 674–78.

Mauras, N. and B. Beaufrere. "Recombinant Human Insulin-Like Growth Factor–1 Enhances Whole Body Protein Anabolism and Significantly Diminishes the Protein Catabolic Effects of Prednisone in Humans Without a Diabetogenic Effect." *Journal of Endocrinology and Metabolism* 80, no. 3 (1995): 869–74.

Mornex, R., C. Jaffiol, and J. Leclere, Eds. *Progress in Endocrinology: The Proceedings of the Ninth International Congress of Endocrinology.* Lancaster, UK; Nice, France: Parthenon Publishing Group, 1992.

Patchett, A.A., et al. "Design and Biological Activities of L–163, 191 (MK–0677): A Potent, Orally Active Growth Hormone Secretagogue." *Medical Science* 92 (July 1995): 7001–05.

Penalva, A., et al. "Effect of Growth Hormone (GH)-Releasing Hormone (GHRH), Atropine, Pyridostigmine, or Hypoglycemia on GHRP–6-Induced GH Secretion in Man." *Journal of Clinical Endocrinology and Metabolism* 76, no. 1 (1993): 168–71.

Pihoker, Catherine, Rosalyn Middleton, George Ann Reynolds, et al. "Diagnostic Studies with Intravenous and Intranasal Growth Hormone-Releasing Peptide–2 in Children of Short Stature." *Journal of Clinical Endocrinology and Metabolism* 80, no. 10 (1995): 2987–92.

Pong, S-S., Lee-Yuh P. Chaung, Dennis C. Dean, et al. "Identification of a New G-Protein-Linked Receptor for Growth Hormone Secretagogues." *Molecular Endocrinology* 10, no. 1 (1996): 57–61.

Popovic, Vera, Svetozar Damjanovic, Dragan Micic, et al. "Growth Hormone (GH) Secretion in Active Acromegaly After the Combined Administration of GH-Releasing Hormone and GH-Releasing Peptide–6." *Journal of Clinical Endocrinology and Metabolism* 79, no. 2 (1994): 456–60.

Rosen, C.J. "Insulin-Like Growth Factors and Bone: The Osteoporosis Connection." *Proceedings of the Society for Experimental Biology* 206 (1994): 83–102.

Sassolas, G., et al. "Clinical Studies with Human Growth Hormone Releasing Factor in Normal Adults and Patients." *Peptides* suppl. 1 (1986): 281–86.

Smith, R.G., et al. "Modulation of Pulsatile GH Release Through a Novel Receptor in Hypothalamus and Pituitary Gland." *Journal of Molecular Endocrinology* 1, no. 10 (1996): 261–86.

Smith, Roy, G., Kang Cheng, William R. Schoen, et al. "A Nonpeptidyl Growth Hormone Secretagogue." *Science* 260 (June 11, 1993): 1640–43.

Thorner, M.O., et al. "Growth Hormone-Releasing Hormone and Growth Hormone-Releasing Peptide As Potential Therapeutic Modalities." *Acta Padiatr. Scand.* 367, Suppl. (1990): 29–32.

Tuilpakov, A.N., et al. "Growth Hormone (GH)-Releasing Effects of Synthetic Peptide GH-Releasing Peptide–2 and GH-Releasing Hormone in Children with GH Insufficiency and Idiopathic Short Stature." *Metabolism* 44, no. 9 (September 1996): 1199–1204.

Waldholz, M. "Merck Develops Type of Human Growth Drug." *Wall Street Journal*, June 11, 1993.

Index